RUBRICS
for Assessing Student Achievement
in SCIENCE GRADES K-12

RUBRICS
for Assessing Student Achievement
in SCIENCE GRADES K-12

HAYS B. LANTZ, Jr.   Foreword by Jay McTighe

CORWIN PRESS
A Sage Publications Company
Thousand Oaks, California

For information:

Corwin Press
A Sage Publications Company
2455 Teller Road
Thousand Oaks, California 91320
www.corwinpress.com

Sage Publications Ltd
1 Oliver's Yard
55 City Road
London EC1Y 1SP
United Kingdom

Sage Publications India Pvt. Ltd.
B-42, Panchsheel Enclave
Post Box 4109
New Delhi 110 017 India

Printed in the United States of America

This book is printed on acid-free paper.

04   05   06   07   08   10   9   8   7   6   5   4   3   2   1

Acquisitions editor:	Faye Zucker
Editorial assistant:	Stacy Wagner
Production editor:	Sanford Robinson
Copy editor:	Liann Lech
Typesetter:	C&M Digitals (P) Ltd.
Proofreader:	Eileen Delaney
Cover designer:	Michael Dubowe
Indexer:	Karen McKenzie

Contents

Foreword

What does it mean to be "standards based" in science education? Agreement about the content standards—what students should know and be able to do—is but the first step. Science educators must also concur on the evidence needed to confirm that students have achieved the desired results.

Objective tests and quizzes will provide appropriate evidence for those standards and companion benchmarks that focus on factual knowledge. However, when the science standards and benchmarks call for students to demonstrate experimental inquiry, systematic reasoning, procedural skills, and scientific habits of mind, performance-based assessment methods are required.

Because performance assessments are inherently open-ended, they typically do not yield a single, correct answer. Thus, the resulting student responses—products and performances—cannot be scored with an answer key or a Scantron machine. Evaluation of student products and performances must be judgment-based—and observations and judgments are inherently subjective.

To improve the quality of evaluation, judgments need to be guided by explicitly defined criteria. These criteria serve to make an essentially subjective process as clear, consistent, and defensible as possible. We cannot claim to be standards-based in science education if different teachers employ different criteria in judging the degrees of student understanding, proficiency, or quality.

It is in the context of standards-based performance assessment and evaluation that *Rubrics for Assessing Student Achievement in Science K-12* provides such a valuable resource. Combining the clarity and detail of the scholar with the practicality of a veteran educator, H. B. Lantz offers the most complete collection of evaluation tools in science available today.

Several key features distinguish this book. The scoring tools address a wide range of important products and performances found in effective science classrooms and programs. The evaluative criteria are embedded within three useful formats—performance lists, holistic rubrics, and analytic rubrics—to accommodate different assessment purposes (diagnostic, formative, and summative). The assessment tools are differentiated by learning levels. This provides an important "scaffolding" of increasingly complex and sophisticated performance expectations across the grades. Thus, the tools not only support improved classroom assessment, they suggest a coherent spiraling for K-12 science programs. Finally, each of the tools has been extensively field-tested over a number of years in K-12 classrooms.

What I most appreciate about these rubrics is their specificity. Unlike the generic rubrics in wide use for grading purposes, these tools contain clear descriptions of the particular traits and qualities that are desired in student products and performances. This specificity yields three significant benefits: (a) more consistent and defensible judgments; (b) more precise feedback, a necessary condition for improving performance; and (c) sharper learning and performance targets for teachers and students.

In sum, *Rubrics for Assessing Student Achievement in Science K-12* provides practical and proven tools for assessing and improving learning and performance in science. You and your students will unquestionably benefit from their use.

Jay McTighe

Jay McTighe is an educational consultant and coauthor of *Assessing Learning in the Classroom* (NEA, 1995), *Understanding by Design* (ASCD, 1998), and *Scoring Rubrics in the Classroom* (Corwin Press, 2000).

Preface

M^{ost} states have developed their own science education standards and, in numerous cases, have begun to assess student achievement of these standards. These actions are, for the most part, a direct response to the publication of national standards for science education by both the American Association for the Advancement of Science (AAAS, 1993) and the National Research Council (NRC, 1996). Many of these science standards are inquiry-based, robust, and comprehensive, and they require complex responses that do not have one single "best" answer. Therefore, judgment-based evaluations will be necessary to assess many of these standards.

As a result, the need has emerged among science educators for assessment and evaluation tools that will complement and extend traditional selected-response test items. In fact, many classroom teachers have requested such tools for some time as they implement and assess student progress on state-mandated standards. The assessment tools within this packet were designed to begin to address this need. Each has been field-tested and revised for many years with elementary, middle, and high school students in Grades K-12. Science educators are encouraged to use the tools as they currently exist or revise them as needed. Teachers and students alike can use them for both formative (embedded) and summative assessments. Ideally, students should be able to view the tool(s) prior to, during, and after the required task is completed. Then, after the student has completed the task and checked the performance list or rubric, he or she may wish to revisit the task to make needed revisions. Finally, the classroom teacher may want to provide the student with a score or grade, make comments to highlight strengths, and suggest additional improvements. The recording and reporting format of achievement for each tool contained within is consistent with this use.

All of the tools within this publication are generic (non-task specific) in design. This means they can be applied to a wide variety of situations in science education grades K-12 without major modifications being necessary. Although the tools are generic, they can serve also as a foundation for defining the criteria for constructing task-specific tools.

Three different versions of assessment tools are included within this publication: performance lists, holistic rubrics, and analytic rubrics. However, not all three forms are available for every performance. Performance lists are the easiest version for teachers to design and for students to use. In addition, the performance list defines the criteria for the task and, therefore, is the foundation for creating both holistic and analytic rubrics. Each version of the tools is written in language that is appropriate for both students in Grades K-12 and teachers.

Acknowledgments

Many individuals have contributed to this manuscript. To the many elementary, middle, and high school science teachers who used and made suggestions for revisions to the assessment tools within this publication, the author is indebted.

Much appreciation is extended to Kathleen Damonte, Lisa Donmoyer, Jean Kugler, and Connie Flowers, science education leaders in the Prince George's County Public Schools, Maryland, for their contributions to sections on elementary science assessment tools. The author will always be grateful for their insights, wit, humor, and professionalism. To Jay McTighe, former director of the Maryland Assessment Consortium and coauthor with Grant Wiggins of *Understanding by Design,* many thanks are given. Jay not only wrote the foreword, but also provided much inspiration to bring this manuscript to publication. Jay's friendship has been a valued commodity throughout much of the author's career.

And to my wife Nancy, a former earth science teacher, who lovingly provided encouragement and expert review, a great debt of gratitude is owed.

Corwin Press extends its thanks to the following reviewers for their contributions to this book:

Sandra K. Enger, The University of Alabama in Huntsville, Huntsville, AL

Diane Holben, Saucon Valley High School, Hellertown, PA

Elizabeth Lolli, Mayfield City Schools, Stow, OH

Douglas Llewellyn, Penfield, NY

Deborah Trumbull, Cornell University, Ithaca, NY

About the Author

Hays B. Lantz, Jr., Ed.D., is currently Director of Science PreK-12 for the Baltimore County Public Schools, Maryland, the 23rd largest school district in the United States. Prior to this assignment he was Supervisor of Science K-12 for the Prince George's County Public Schools, Maryland, Supervisor of the Howard B. Owens Science Center in Prince George's County, Federal Programs Director, Director of Curriculum and Instruction, Title IV-C Project Director, and high school administrator and classroom science teacher. Dr. Lantz has either taught or supervised all grades K-16 and has been extensively involved in professional and program development in science education for 34 years. He has successfully directed several National Science Foundation projects, most recently the Partnerships for Math and Science Achievement (PMSA). In addition, Dr. Lantz also stays active as a science education consultant for several publishers and numerous school districts. His current research interests in science education include performance-based teaching and assessment, advance organizers, and the effects of prior knowledge on learning.

Dr. Lantz received his formal education at James Madison University and at the University of Virginia. He earned his doctorate in Science Education from the University of Virginia in 1981, with supporting fields in Research & Evaluation and Supervision & Administration.

When not actively engaged in his work, Dr. Lantz enjoys skiing, golfing, fly fishing, hiking, and other outdoor activities.

**CORWIN
PRESS**

The Corwin Press logo—a raven striding across an open book—represents the union of courage and learning. Corwin Press is committed to improving education for all learners by publishing books and other professional development resources for those serving the field of K–12 education. By providing practical, hands-on materials, Corwin Press continues to carry out the promise of its motto: **"Helping Educators Do Their Work Better."**

Chapter 1

Assessment in Science Education

A Call for Change

There is much agreement within the science education community that current methods for assessing student achievement, primarily selected response tests, fall short of measuring standards that are part of contemporary science education. Most testing efforts, including standardized, norm-referenced, and teacher-designed tests, are not consistent with national and state standards and benchmarks. Some critics have gone so far as to charge that traditional assessments have been isolated from, and consequently are actually damaging to, instruction and learning (Jamentz, 1994). Also, many others contend that traditional assessments are merely used to sort students and, as a result, may deny educational opportunities (Darling-Hammond, 1991). If indeed these claims are the case, and if standard written tests are no longer adequate for measuring a student's performance on the standards of science instruction, then what kinds of alternative assessments are needed? What guiding philosophies and principles will be used to shape these new assessments? What formats will these assessments take? Who will design them?

Current thinking is that science instruction must be evaluated using quite different techniques and tools. Techniques are needed that show not only what students know, but also how students actually use and apply information and skills. If we want students to solve problems, answer open-ended questions, and actually perform as is called for by many educational reformers (Silver, Strong, & Perini, 2000), then tests must be developed that measure performance. Most traditional paper-and-pencil tests that evaluate answers as being either right or wrong cannot adequately evaluate performance. Think about the work of an architect, an engineer, a musician, a teacher, a baseball player, or a writer. Either a performance or a product is the basis for evaluating their work. These people do not take paper-and-pencil tests to demonstrate what they know—they perform! The question "Why don't we apply these same ideas to teaching and assessing student achievement?" naturally arises at this point.

When students are assessed based upon their performance, then constant feedback will become an integral part of the instructional process. If we take this approach, assessment becomes a process and not an event. A coach works constantly with his or her players to improve upon their performance. The same approach can and should apply to teaching science. The sequence of teach, assess, and adjust instruction based upon the results of assessment is a methodology that works to improve performance. Think back to a class you took in college. Didn't you typically do your best work when you were allowed to develop a draft, get feedback from the professor, and then make revisions before turning in the final product?

USING NATIONAL AND STATE STANDARDS TO IMPLEMENT CHANGE

National science standards from the American Association for the Advancement of Science (1993) and the National Research Council (1996) have been part of the science education environment for about 10 years. More recently, state-specific science standards have come to dominate the science curriculum for growing numbers of students. But this movement is not without controversy, because some have criticized state standards as being inadequate (Lerner, 1998). And teachers, administrators, students, and parents alike continue to question the need for such standards. Standards for some seem unnecessarily burdensome to implement, seem restrictive to the curriculum, and lead to increased testing. Their skepticism is somewhat justified in light of the numerous supposedly "innovative practices" that have been perpetuated upon them in the past and that, in many cases, were based upon nothing more than someone's good idea, without any real supporting evidence from field studies.

It has become increasingly evident that if science educators are ever going to enthusiastically embrace these standards, then they must have opportunities to "unpackage" standards by interpreting their intent and by designing units of study. In so doing, standards will become "real" and useful to them, especially when they continue to learn through gathering evidence from students that a standard has been met or progress is being made toward meeting it.

The encouraging news is that the continued use of standards and assessment of standards in educational settings is leading many educators to believe that standards can indeed

- Lead to improved student performance (Andrade, 1997; Marzano, 2001)
- Motivate educators to explore questions at the very heart of the purposes and processes of schooling (Jamentz, 1994)
- Provide both students and teachers with developmentally appropriate benchmarks of student performance that can be used to design performance tasks and performance-based units
- Provide both students and teachers with well-defined criteria by which to assess student performance and understanding
- Provide the foundation for the development of a focused professional development program for all stakeholders
- Provide a stable foundation upon which instructional programs of excellence can be constructed and maintained year after year (Lantz, 2001)
- Provide for quality control and consistency of teaching, assessment, and accountability across classrooms, schools, and school districts
- Provide ample opportunities for students to become more thoughtful judges of the quality of their own work and to revisit past performances so as to improve upon them
- Reduce the amount of time teachers spend evaluating student work, as a result of more student self- and peer review

Today's science standards are comprehensive in skills and processes, inquiry, and science content; are robust and rich; often have multiple "right" answers; and require performances to assess them. Consequently, traditional modes of assessment alone are not sufficient to gather evidence of student understanding of these standards. As a result, complementary and alternative forms of assessment have emerged. Alternative assessment means any assessment format that is nontraditional and requires the student to construct, demonstrate, or perform (Doran, Chan, & Tamir, 1998).

ASSESSMENT TECHNIQUES

When it comes to assessment, there exists a good news-bad news scenario. The good news is that classroom teachers already use a variety of assessment techniques that could be used to evaluate performance. The bad news is that these techniques are generally thought of as informal assessment methods and are rarely used for formal grading or adjustment of instruction.

In the science classroom, any behavior that can be observed can be assessed. Observing what students write, say, and do can form the foundation for assessing performance. Projects, interviews, conferences, presentations, journals, logs of data and observations, lab reports, extended studies, and student self-assessments are all performances that can provide evidence of student understanding or lack thereof. Unfortunately, the power of using assessment of these types of performances for formative feedback and adjustment of instruction still remains unrealized in many science classrooms of today. Here is where the concept of the performance task takes center stage.

PERFORMANCE TASKS

The purpose of a performance task is to assess what students know and what they can do with what they know. The use of performance tasks to assess student achievement, particularly science process skills, has been well documented. Their reliability and validity parallels that of traditional assessments (Adams & Callahan, 1995). Science performance tasks take many formats depending upon their intended audience and purpose. However, according to Wiggins and McTighe in *Understanding by Design* (1998), performance tasks should consist of complex challenges that reflect problems and issues faced by adults. In addition, the task should be meaningful, authentic, and worth mastering. Typically, the student knows in advance the goal, the role, the audience, the setting, the product or performance, and the standards against which work will be assessed (McTighe, 1999). Performance tasks, when used as summative assessments for a unit of study, serve to hook the student, activate prior knowledge, and let the student know where the unit is headed.

The following criteria are often used to shape performance tasks. A science performance task should

- Provide opportunities for students to demonstrate and communicate their understanding of standards, benchmarks, goals, objectives, and science content (Baron, 1991)
- Serve to anchor the unit, lesson, or performance (Wiggins & McTighe, 1998)
- Afford students an opportunity to demonstrate their *understanding* of science and not just provide a single, best, and often superficial answer
- Integrate knowledge and skills within the disciplines of science, language arts, and mathematics
- Be meaningful, authentic, interesting, challenging, and thought-provoking, and have cognitively appropriate content (Baron, 1991)
- Stress depth over breadth
- Allow for multiple approaches, solutions, and answers and not have one clear path of action specified at the beginning of the task
- Raise other questions or lead to other problems

ASSESSING STUDENT ACHIEVEMENT ON PERFORMANCE TASKS

Performance tasks, by their very nature, cannot be evaluated using traditional paper-and-pencil tests. Performance tasks involve understanding of scientific concepts and procedures; stimulate depth of thought, are usually open-ended, and seldom have one correct answer. The evaluation of such tasks involves the development of assessment tools and the professional judgment of trained educators. Before challenging students to attempt a performance, the teacher should explain the task, give the students written criteria, discuss the criteria, and provide examples of exemplary performance (anchors). Criteria should always be in writing and well understood by everyone before student performances begin.

The first step in evaluating a performance task is to establish a system of documenting students' performances. Two assessment tools—the Performance List Rubric (Chapter 2) and the Holistic and/or Analytic Rubric (Chapter 3)—are presented within this text. Both are versatile and effective techniques. They are not merely abstract numbering systems but instead are a taxonomic system that provides specific assessment guidelines for teachers and students alike. Both can provide not only summative but also formative assessments.

To be used effectively, these two assessment tools must be used at various stages throughout the performance. This gives students a chance to perform, receive feedback, and revise their work. Thus, the performance list rubric and the holistic and/or analytic rubric allow students and teachers to establish a classroom environment in which risk taking and creativity are rewarded. Students are given credit for making incremental progress toward a goal and are not penalized for a lack of immediate success (Treagust, Jacobowitz, Gallagher, & Parker, 2003).

Assessment tools can be general (non-task specific), or they can be developed to measure student progress toward mastery of specific skills and knowledge. Whatever tool (specific or non-task specific) is used, criteria must be clear and unambiguous so that students know what performance is needed to reach educational goals.

Performance assessments allow students to compete with themselves, rather than with other students. Through such assessments, students can gain a real understanding of what they know and what they can do. Performance assessments, unlike written tests, need not be threatening. Because there are many correct answers, performance assessments can take the fear out of learning science. Taking the fear and anxiety out of the science classroom may motivate many more students to continue their study of science. They will also enjoy, learn, and use more science. Performance assessment makes school learning more relevant to students' lives and the real world. It helps teachers focus on the really important outcomes of education, instead of teaching isolated bits of information. As students learn to become competent problem solvers and to be confident of their ability to think logically and communicate their ideas clearly, they will recognize that they have received an education that has prepared them for life—for life as productive citizens in the 21st century.

What Might a Performance List Rubric Look Like for a Second-Grade Performance Task?

Example 1. Signals of Spring

Ms. Logan, principal of Mother Jones Elementary School, wants to plant 1,000 tulips this fall so that they will be ready to bloom in the spring. Like all plants, tulips need air, water, and food to live. Plants get most of their water from the soil in which they are planted. Before the tulips can be planted, you need to find out if the soil is the right kind of soil for the tulips. You will test the soil at Mother Jones as well as other kinds of soil to see which would be the best for growing tulips. Then you will communicate your results to Ms. Logan through scientific drawings and a report to tell her how to best prepare the soil for the tulips.

Example 1 is a performance task that has been used successfully with elementary students in Grade 2. This task was designed to address the following national standards:

Project 2061 Benchmarks

Everyone can do science and invent things and ideas. (Chapter 1C, The Nature of Science: Grades K-2)

People can often learn about things around them by just observing those things carefully, but sometimes they can learn more by doing something to things and noting what happens. (Chapter 1B, The Nature of Science: Grades K-2)

Plants and animals have features that help them live in different environments. (Chapter 5, The Living Environment: Grades K-2)

Most living things need water, food, and air. (Chapter 5, The Living Environment: Grades K-2)

National Science Education Standards

Scientific investigations involve asking and answering a question and comparing the answer with what scientists already know about the world. (Content Standard A: Science as Inquiry: Grades K-4)

Scientists make the results of their investigations public; they describe the investigation in ways that enable others to repeat the investigations. (Content Standard A: Science as Inquiry: Grades K-4)

Organisms have basic needs. For example, animals need air, water, and food; plants require air, water, nutrients, and light. Organisms can survive only in environments in which their needs can be met. The world has many different environments, and distinct environments support the life of different types of organisms. (Content Standard C: Life Science: Grades K-4)

The performance list rubric that was used to assess the scientific drawings of the students in Example 1 appears in Figure 1.1.

The columns labeled "Self" and "Teacher" can be used in several ways. Checkmarks (✓) could be placed in the column "Self" if the student believes the criterion is present in his or her performance. The teacher can then add checks in the "Teacher" column to validate and reinforce the student's self-assessment. Or each criterion (1-4) could be weighted by designating points for each, and then "Self" and/or "Teacher" assessments can be completed.

What Might a Performance List Rubric Look Like for a Fourth-Grade Performance Task?

Example 2. Crumbling Monuments

Have you ever heard of acid rain? You may have heard newscasters talking about it on television or read about it in a newspaper or magazine. What exactly is acid rain, and how harmful is it? How does it affect plants, animals, buildings, and people? The Capitol building is a national monument that is being affected by acid rain.

In order to understand acid rain, you need to know what acids are. In this unit, you will first explore acids and bases and how scientists identify them. You will then learn about acid rain. At the end of the unit, you will create a PowerPoint presentation on acid rain that could be shown during morning announcements at your school. The purpose of your presentation is to make students aware of the problem of acid rain. You may want to review the performance list rubric for "PowerPoint Presentation" before you begin.

Example 2 is a performance task that has been used successfully with elementary students in Grade 4. This task was designed to address the following national standards:

Project 2061 Benchmarks

Technologies often have drawbacks as well as benefits. A technology that helps some people or organisms may hurt others—either deliberately (as weapons can) or inadvertently (as pesticides can). When harm occurs or seems likely, choices have to be made or new solutions found. (Chapter 3C, Issues in Technology: Grades 3-5)

When liquid water disappears, it turns into a gas (vapor) in the air and can reappear as a liquid when cooled, or as a solid if cooled below the freezing point of water. Clouds and fog are made of tiny droplets of water. (Chapter 4B, The Physical Setting: Grades 3-5)

Scientific Drawing

Name _____

Date _____

Topic _____

	Performance Criteria	Assessment		
		Points	Self	Teacher
1.	My scientific drawing shows details of what I actually observed.			
2.	All parts of my scientific drawing are clearly and accurately labeled.			
3.	My drawing has a title that explains what the drawing is all about.			
4.	My scientific drawing is large enough to see all parts clearly.			

Teacher Comments:

Figure 1.1   An Example of a Performance List Rubric for a Scientific Drawing for Grades 2-3

For any particular environment, some kinds of plants and animals survive well, some survive less well, and some cannot survive at all. (Chapter 5D, Interdependence of Life: Grades 3-5)

National Science Education Standards

Scientific investigations involve asking and answering a question and comparing the answer with what scientists already know about the world. (Content Standard A: Science as Inquiry: Grades K-4)

Scientists develop explanations using observations (evidence) and what they already know about the world (scientific knowledge). Good explanations are based on evidence from investigations. (Content Standard A: Science as Inquiry: Grades K-4)

Materials can exist in different states—solid, liquid, and gas. Some common materials, such as water, can be changed from one state to another by heating or cooling. (Content Standard B: Physical Science: Grades K-4)

Organisms have basic needs. For example, animals need air, water, and food; plants require air, water, nutrients, and light. Organisms can survive only in environments in which their needs can be met. The world has many different environments, and distinct environments support the life of different types of organisms. (Content Standard C: Life Science: Grades K-4)

Earth materials are solid rocks and soils, water, and the gases of the atmosphere. The varied materials have different physical and chemical properties, which make them useful in different ways, for example, as building materials, as sources of fuel, or for growing the plants we use as food. Earth materials provide many of the resources that humans use. (Content Standard D: Earth and Space Science: Grades K-4)

The performance list rubric that was used to assess the PowerPoint Presentation of the students in Example 2 appears in Figure 1.2.

What Might a Performance List Rubric and a Holistic Rubric Look Like for a Middle School (Grade 7) Performance Task?

Example 3. Get Into the Swing

For generations, your family has owned an antique grandfather clock. Recently, your grandparents gave the clock to your family as a present. However, during transport of the clock, the pendulum became detached. The pendulum has been reattached and the clock has been set up at your home, but it no longer keeps accurate time. Members of your family all have different opinions about how to adjust the pendulum so the clock will keep accurate time again. Your sister believes that if you adjust the length of the pendulum, you will be able to correct the problem. Your father believes that if you change the mass of the pendulum, you can increase its accuracy. Your mother believes that if you adjust the amplitude (the amount of displacement of the pendulum from its resting position), you can correct the problem. Your grandparents are quite concerned. What do you think?

You need to communicate with your grandparents, advising them of your solution. One format you might want to consider is a friendly letter. If you choose this format, you might want to review the performance list rubric and/or the holistic rubric "Writing to Inform" before you begin. Regardless of how you communicate

PowerPoint Presentation

Name _____   Date _____   Course/Class _____

Task/Assignment _____

	Assessment			
Performance Criteria	**Points**	**Self**	**Teacher**	**Other(s)**
1. The topic has been extensively and accurately researched.				
2. A storyboard, consisting of logically and sequentially numbered slides, has been developed.				
3. The introduction is interesting and engages the audience.				
4. The fonts are easy to read and point size varies appropriately for headings and text.				
5. The use of italics, bold, and underline contributes to the readability of the text.				
6. The background and colors enhance the text.				
7. The graphics, animations, and sounds enhance the overall presentation.				
8. Graphics are of proper size.				
9. The text is free of spelling, punctuation, capitalization, and grammatical errors.				

Comments	Goals	Actions

Figure 1.2   An Example of a Performance List Rubric for PowerPoint Presentation for Grades 4-6

your results, it will be evaluated based upon how well you organize and develop the topic using appropriate and scientifically correct language.

Example 3 is a performance task that has been used successfully with seventh-grade students at the end of an integrated science unit on force and motion, gravity and microgravity, and pendulums. This task was designed to address the following national standards:

Project 2061 Benchmarks

Everything on or anywhere near the earth is pulled toward the earth's center by gravitational force. (Chapter 4, The Physical Setting: Grades 6-8)

National Science Education Standards

The motion of an object can be described by its position, direction of motion, and speed. That motion can be measured and represented on a graph. (Content Standard B: Physical Science: Grades 5-8)

The performance list and holistic rubrics that were used to assess student work in example #3 appear in Figures 1.3 and 1.4.

What Might a Performance List Rubric and an Analytic Rubric Look Like for an Earth Science Performance Task?

Example 4. Cruising Contours

Many people love winter sports, and sledding is a popular one. On a cold, snowy day, all that is needed is something to slide on and a hill with a steep pitch. Then, let gravity do the work and enjoy the ride. You may know where there are sled runs in your neighborhood, but most hills in Prince George's County, Maryland, are not steep or long. Did you ever wonder why?

Snowboarding and downhill skiing are sports that you as a teenager really enjoy. However, they require slopes much steeper than the hills found locally. There are no ski resorts in Prince George's County. Are there ski resorts anywhere in the state of Maryland?

Your task is to research the location of existing ski resorts in Maryland and apply what you have learned about landforms, topography, natural resources, and heat and sunlight to explain why the resorts were built there. In addition, you are to propose and defend the location of another future ski resort in Maryland, using topographic maps to fully explain your decision. You are to develop a presentation (you decide the format) for the local Chamber of Commerce of the county in which you wish to propose your new skiing/snowboarding area. Your presentation will be evaluated based upon its clarity, organization, use of language, correct use of scientific concepts, supporting data, and persuasiveness.

Example 3 is a performance task that has been used successfully with earth science students at the end of an integrated unit on landforms, topography, heat and sunlight, and natural resources. This task was designed to address the following national standards:

National Science Education Standards

Landforms are the result of a combination of constructive and destructive forces. Constructive forces include crustal deformation, volcanic eruption, and deposition of sediment, while destructive forces include weathering and erosion. (Content Standard D: Earth and Space Science: Grades 5-8)

Writing to Inform in Science

Name _____ Date _____ Course/Class _____

Task/Assignment _____

Performance Criteria

	Assessment		
Points	**Self**	**Teacher**	**Other(s)**

1. Accurate, specific, and purposeful scientific facts and concepts are included and are extended and expanded to fully explain the topic.

2. A logical organizational plan for the text is established and consistently maintained.

3. Scientific information that is relevant to the needs of the audience is used throughout the text.

4. Scientific vocabulary and language choices enhance the text.

5. Diagrams, pictures, and other graphics are of quality and add to the overall effectiveness of the text.

6. There are no errors in the mechanics (spelling and grammar).

Comments   Goals   Actions

Figure 1.3   An Example of a Performance List Rubric for Writing to Inform for Grades 7-12

Writing to Inform in Science

Name _____ Date _____ Course/Class _____

Task/Assignment _____

Expert 4	<u>Development</u>: The writer provides accurate, specific, and purposeful scientific facts and concepts that are extended and expanded to fully explain the topic. <u>Organization</u>: The writer establishes an organizational plan and consistently maintains it. <u>Audience</u>: The writer provides scientific information relevant to the needs of the audience. <u>Language</u>: The writer consistently provides scientific vocabulary and language choices to enhance the text. There are no errors in the mechanics (spelling and grammar).
Proficient 3	<u>Development</u>: The writer provides scientific facts and concepts that adequately explain the topic with some extension of ideas. The information is usually accurate and purposeful. <u>Organization</u>: The writer establishes and maintains an organizational plan, but the plan may have some minor flaws. <u>Audience</u>: The writer provides information most of which is relevant to the needs of the audience. <u>Language</u>: The writer frequently provides scientific vocabulary and uses language choices to enhance the text. There are few errors in the mechanics (spelling and grammar).
Emergent 2	<u>Development</u>: The writer provides scientific facts and concepts that inadequately explain the topic. The information is sometimes inaccurate, general, or extraneous. <u>Organization</u>: The writer generally establishes and maintains an organizational plan. <u>Audience</u>: The writer provides some information relevant to the needs of the audience. <u>Language</u>: The writer sometimes provides scientific vocabulary and uses language choices to enhance the text. There are significant errors in the mechanics (spelling and grammar).
Novice 1	<u>Development</u>: The writer provides insufficient scientific facts and concepts to explain the topic. The information provided may be vague or inaccurate. <u>Organization</u>: The writer either did not establish an organizational plan or, if an organizational plan is established, it is only minimally maintained. <u>Audience</u>: The writer did not provide information relevant to the needs of the audience. <u>Language</u>: The writer seldom, if ever, provides scientific vocabulary and uses language choices to enhance the text. There are many errors in the mechanics (spelling and grammar).

Comments	Goals	Actions

Figure 1.4   An Example of a Holistic Rubric for Writing to Inform for Grades 7-12

The sun is a major source of energy for changes on the earth's surface. The sun loses energy by emitting light. A tiny fraction of that light reaches the earth, transferring energy from the sun to the earth. The sun's energy arrives as light with a range of wavelengths, consisting of visible light, infrared, and ultraviolet radiation. (Content Standard B: Physical Science: Grades 5-8)

The performance list and analytic rubrics that were used to assess student work appear in Figures 1.5 and 1.6. The two small boxes in each cell of the analytic rubric could be used for placing checkmarks (✓) for self-, peer, teacher, and/or other assessments.

What Might a Performance List Rubric Look Like for a High School Chemistry Performance Task?

Example 5. Warming Up to Chemistry

For years, your extended family has taken a skiing/snowboarding trip to New England. You usually hang out with your cousin, who is the same age as you and who is also taking chemistry. There's only one problem: You have to stop often to go inside and warm up because your cousin's hands and feet get cold. You want to spend time with your cousin this year as well, but you don't want to have to stop all the time. You decide to do a little research and investigate the "toe heaters" and "hand warmers" people have told you about. Once you've finished your investigation, you will write a letter to your cousin explaining how the hot packs work and why he should consider using them. You will also include other suggestions about keeping warm in winter. Your cousin is a real science buff and fellow chemistry student, so you'll have to explain the scientific concepts completely, accurately, and in an organized fashion.

Example 5 was a performance task used with students in chemistry at the end of a unit on thermochemistry. The following national standards are addressed within this unit:

Project 2061 Benchmarks

Transformations of energy usually produce some energy in the form of heat, which spreads around by radiation or conduction into cooler places. Although just as much total energy remains, its being spread out more evenly means less can be done with it. (Chapter 4, The Physical Setting: Grades 9-12)

The rate of reactions among atoms and molecules depends on how often they encounter one another, which is affected by the concentration, pressure, and temperature of the reacting materials. Some atoms and molecules are highly effective in encouraging the interaction of others. (Chapter 4, The Physical Setting: Grades 9-12)

National Science Education Standards

Chemical reactions may release or consume energy. Some reactions such as the burning of fossil fuels release large amounts of energy by losing heat and by emitting light. Light can initiate many chemical reactions such as photosynthesis and evolution of urban smog. (Content Standard B: Physical Science: Grades 9-12)

Chemical reactions can take place in time periods ranging from the few femtoseconds (10–15 seconds) required for an atom to move a fraction of a chemical bond distance to geologic time scales of billions of years. Reaction rates depend on how often the reacting atoms and molecules encounter one another, on the temperature, and on the properties—including shape—of the reacting species. (Content Standard B: Physical Science: Grades 9-12)

The task-specific holistic rubric used to assess student work in Example 5 appears in Figure 1.7.

Writing to Persuade in Science

Name _____   Date _____   Course/Class _____

Task/Assignment _____

Performance Criteria	Assessment			
	Points	Self	Teacher	Other(s)
1. A clear position is established that is fully supported or refuted with relevant, accurate scientific and/or personal information.				
2. A logical organizational plan for the text is established and consistently maintained.				
3. Scientific information that is relevant to the needs of the audience is used throughout the text.				
4. Scientific vocabulary and language choices enhance the position.				
5. Diagrams, pictures, and other graphics are of quality and add to the overall effectiveness of the position.				
6. There are no errors in the mechanics (spelling and grammar).				

Comments	Goals	Actions

Figure 1.5   An Example of a Performance List Rubric for Writing to Persuade for Grades 7-12

Name _____

Task/Assignment _____

Date _____   Course/Class _____

Writing to Persuade in Science (Analytic Rubric)

	Development	Organization	Audience	Language
Expert **4**	Development: The writer identifies a clear position and fully supports or refutes that position with relevant, accurate scientific and/or personal information. ☐	Organization: The writer presents an organizational plan that is logical and consistently maintained. ☐	Audience: The writer effectively addresses the needs and characteristics of the identified audience. ☐	Language: The writer consistently uses relevant, scientific vocabulary and language choices to enhance the text. ☐
Proficient **3**	Development: The writer identifies a clear position and partially supports or refutes that position with relevant, accurate scientific and/or personal information. ☐	Organization: The writer presents an organizational plan that is logical and maintained, but with minor flaws. ☐	Audience: The writer adequately addresses the needs and characteristics of the identified audience. ☐	Language: The writer frequently uses relevant, scientific vocabulary and language choices to enhance the text. ☐
Emergent **2**	Development: The writer identifies a position, yet that position lacks clarity. The writer tries to support or refute that position with relevant, accurate scientific and/or personal information. ☐	Organization: The writer presents an organizational plan that is only generally maintained. ☐	Audience: The writer minimally addresses the needs and characteristics of the identified audience. ☐	Language: The writer sometimes uses scientific vocabulary and language choices to enhance the text. ☐
Novice **1**	Development: The writer identifies an ambiguous position with little or no relevant, accurate scientific and/or personal information to support that position; or the writer fails to identify a position. ☐	Organization: The writer presents an argument that is illogical and/or minimally maintained. ☐	Audience: The writer does not address the needs and characteristics of the identified audience. ☐	Language: The writer seldom, if ever, uses scientific vocabulary and language choices to enhance the text. ☐

Figure 1.6   An Example of an Analytic Rubric for Writing to Persuade for Grades 7–12

Task-Specific Holistic Rubric for "Warming Up to Chemistry"

Name _____ Date _____ Course/Class _____

Task/Assignment _____

Expert 4	The following concepts are completely and accurately explained in the letter: heat and temperature and their differences; kinetic energy and temperature; heat transfer and measurement; endothermic and exothermic reactions; positive and negative enthalpy and surroundings; Hess's Law; enthalpy and entropy; and Gibbs free energy. The response includes a recommendation for a hot pack that is extensively supported, along with its availability and cost analysis.
Proficient 3	The following concepts are mostly explained in the letter: heat and temperature and their differences; kinetic energy and temperature; heat transfer and measurement; endothermic and exothermic reactions; positive and negative enthalpy and surroundings; Hess's Law; enthalpy and entropy; and Gibbs free energy. The response includes a recommendation for a hot pack that is supported, along with its availability and cost analysis.
Emergent 2	The following concepts are partially explained in the letter: heat and temperature and their differences; kinetic energy and temperature; heat transfer and measurement; endothermic and exothermic reactions; positive and negative enthalpy and surroundings; Hess's Law; enthalpy and entropy; and Gibbs free energy. The response includes a recommendation for a hot pack, but without being supported. Availability and cost analysis of the hot pack may or may not be included.
Novice 1	The following concepts are explained in the letter, but contain many misunderstandings and misconceptions: heat and temperature and their differences; kinetic energy and temperature; heat transfer and measurement; endothermic and exothermic reactions; positive and negative enthalpy and surroundings; Hess's Law; enthalpy and entropy; and Gibbs free energy. The response may or may not include a recommendation for a hot pack

Comments	Goals	Actions

Figure 1.7 An Example of a Task-Specific Holistic Rubric for Chemistry

What Might a Performance List Rubric Look Like for a High School Biology Performance Task?

Example 6. The Great Divide

A local middle school science teacher, Ms. Jones, wants to start a mentoring program in science for her seventh-grade classes and wants high school students to be the mentors. Middle school students are most impressed by high school students, so your impact upon them could be tremendous. The first topic Ms. Jones wants you to teach the middle school students is—you guessed it—cell division, including mitosis and meiosis! But it has been several years since you last studied cell division, so you know you need to brush up. As a result, your biology teacher has developed the following unit for you to refresh your memory. At the end of this unit, you must demonstrate that you understand mitosis and meiosis well enough to teach it to seventh-grade students. To do this, you must be able to develop a poster comparing mitosis and meiosis that you can use to teach the middle school students. You will also prepare a clear written explanation of mitosis and meiosis. Are you ready?

Example 6 was a performance task used with students in biology at the end of a unit on cell division. The following national standards are addressed within this unit:

Project 2061 Benchmarks

Scientists assume that the universe is a vast single system in which the basic rules are the same everywhere. The rules may range from very simple to extremely complex, but scientists operate on the belief that the rules can be discovered by careful, systematic study. (Chapter 1A: The Scientific World View: Grades 9-12)

The degree of kinship between organisms or species can be estimated from the similarity of their DNA sequences, which often closely match their classification based on anatomical similarities. (Chapter 5A: Diversity of Life: Grades 9-12).

The information passed from parents to offspring is coded in DNA molecules. (Chapter 5B: Heredity: Grades 9-12)

National Science Education Standards

Cells store and use information to guide their functions. The genetic information stored in DNA is used to direct the synthesis of the thousands of proteins that each cell requires. (Content Standard C: Life Science: Grades 9-12)

Most of the cells in a human being contain two copies of each of the 22 different chromosomes. In addition, there is a pair of chromosomes that determines sex: a female contains two X chromosomes and a male contains one X and one Y chromosome. Transmission of genetic information to offspring occurs through egg and sperm cells that contain only one representative from each chromosome pair. An egg and a sperm unite to form a new individual. The fact that the human body is formed from cells that contain two copies of each chromosome—and therefore two copies of each gene—explains many features of human heredity, such as how variations that are hidden in one generation can be expressed in the next. (Content Standard C: Life Science: Grades 9-12)

Changes in DNA (mutations) occur spontaneously at low rates. Some of these changes make no difference to the organism, whereas others can change cells and organisms. Only mutations in germ cells can create the variation that changes an organism's offspring. (Content Standard C: Life Science: Grades 9-12)

The rubric used to assess student work in Example 6 appears in Figure 1.8.

What Might a Performance List Rubric Look Like for a High School Physics Performance Task?

Example 7. Charge! On the Move

Welcome to Mr. Covert's Physics Challenge of the Month. Your challenge is to purchase and install a new kick-bass car stereo system in your car. You have a budget of $2,000 to spend on your equipment. Using your knowledge of series and parallel circuits and Ohm's Law, you will have to plan and design the best way to connect the speakers to the amplifier to maximize the power output of the amplifier, but not burn out the amp. You will also need to include a cost analysis for the equipment purchased. Note: You must have at least four 8-ohm speakers and only one amplifier to run the system. The choice of stereo, equalizer, and other equipment is entirely up to you. You must have the following items in this project:

- Cost analysis of equipment to purchase that includes item name, manufacturer of item, place of purchase, and cost
- Schematic diagram of circuit including battery, fuse, radio, equalizer and effects, power amplifier, and speakers
- Calculation of effective resistance of circuit
- Calculation of current drawn from battery and recommended fuse size for the circuit

At the end of your preparation for the challenge, you must present your findings orally in a convincing, clear, and scientifically accurate way. You may use any visual you wish in your presentation.

Example 7 was a performance task used with students in physics at the end of a unit on series and parallel circuits and application of Ohm's Law to various series and parallel circuits.

The following national standards are addressed within this unit. Students discover the properties of series and parallel circuits.

Project 2061 Benchmarks

Scientists assume that the universe is a vast single system in which the basic rules are the same everywhere. The rules may range from very simple to extremely complex, but scientists operate on the belief that the rules can be discovered by careful, systematic study. (Chapter 1A: The Nature of Science: Grades 9-12)

Investigations are conducted for different reasons, including to explore new phenomena, to check on previous results, to test how well a theory predicts, and to compare different theories. (Chapter 1B: Scientific Inquiry: Grades 9-12)

Hypotheses are widely used in science for choosing what data to pay attention to and what additional data to seek, and for guiding the interpretation of the data (both new and previously available). (Chapter 1B: Scientific Inquiry: Grades 9-12)

Poster

Name _____ Date _____ Course/Class _____

Task/Assignment _____

	Assessment			
Performance Criteria	**Points**	**Self**	**Teacher**	**Other(s)**
1. The poster contains a title that clearly reflects the topic or theme.				
2. The poster contains relevant and accurate information about the topic or theme.				
3. The format of the poster is appropriate to the content, purpose, and audience for which it is designed.				
4. Graphic elements, such as pictures, photographs, charts, tables, scientific drawings, diagrams, graphs, etc., add to the overall effectiveness of the poster.				
5. There is a coherent, flowing organization to the poster with the various elements (text, graphics, etc.) working well together.				
6. The poster is aesthetically pleasing, with effective use of space, color, texture, and shape.				
7. The poster is skillfully designed and crafted using appropriate graphic design tools.				
8. The poster effectively communicates its theme in convincing fashion to the intended audience.				
9. The poster is creative and draws attention.				
10. Language chosen for the poster is captivating, persuasive, informative, accurate, and concise.				

Comments	Goals	Actions

Figure 1.8   An Example of a Performance List Rubric for Poster for Grades 7-12

Mathematics provides a precise language for science and technology—to describe objects and events, to characterize relationships between variables, and to argue logically. (Chapter 2B: Mathematics, Science, and Technology: Grades 9-12)

Any mathematical model, graphic or algebraic, is limited in how well it can represent how the world works. The usefulness of a mathematical model for predicting may be limited by uncertainties in measurements, by neglect of some important influences, or by requiring too much computation. (Chapter 9B: Symbolic Relationships: Grades 9-12)

Tables, graphs, and symbols are alternative ways of representing data and relationships that can be translated from one to another. (Chapter 9B: Symbolic Relationships: Grades 9-12)

When a relationship is represented in symbols, numbers can be substituted for all but one of the symbols and the possible value of the remaining symbol computed. Sometimes the relationship may be satisfied by one value, sometimes more than one, and sometimes maybe not at all.

The reasonableness of the result of a computation can be estimated from what the inputs and operations are. (Chapter 9B: Symbolic Relationships: Grades 9-12)

National Science Education Standards

Mathematics is essential in scientific inquiry. Mathematical tools and models guide and improve the posing of questions, gathering data, constructing explanations and communicating results. (Content Standard A: Science as Inquiry: Grades 9-12)

The electric force is a universal force between any two charged objects. Opposite charges attract while like charges repel. The strength of the force is proportional to the charges, and, as with gravitation, inversely proportional to the square of the distance between them. (Content Standard B: Physical Science: Grades 9-12)

In some materials, such as metals, electrons flow easily, whereas in insulating materials such as glass they can hardly flow at all. Semiconducting materials have intermediate behavior. At low temperatures some materials become superconductors and offer no resistance to the flow of electrons. (Content Standard B: Physical Science: Grades 9-12)

The performance list rubric used to assess student work in Example 7 appears in Figure 1.9.

IMPLEMENTING PERFORMANCE TASKS IN THE SCIENCE CLASSROOM

By reviewing the examples above, it is obvious that good performance tasks and assessment tools (performance list rubrics, holistic rubrics, and/or analytic rubrics) take time to construct, field test, and revise. Also, an assessment consisting of rubrics cannot be scored as quickly and as easily as a selected response item. The obvious question that comes to mind at this time is, "Do I, as a very busy classroom teacher, have the time to do this kind of performance evaluation?" The answer is an unequivocal yes. But it may mean that you will have to modify your pedagogy to make your classroom more performance based and student centered, rather than being primarily teacher directed. You may need to meaningfully engage students, provide explorations for data collection and analysis, and give students opportunities to develop and demonstrate

Oral Presentation in Science

Name _____ Date _____ Course/Class _____

Task/Assignment _____

Performance Criteria	Assessment			
	Points	Self	Teacher	Other(s)
Content and Organization 1. The purpose of the presentation (informing, persuading, or both), the subject, and any position taken by the presenter are clearly defined at the outset.				
2. The presentation is made in an interesting, logical sequence – an introduction, an organized body, and a clear closure – that the audience can follow.				
3. The introduction has a strong purpose statement that serves to captivate the audience and narrow the topic.				
4. An abundance of accurate supporting scientific concepts, facts, figures, statistics, scenarios, stories, and analogies are used to support the key points and ideas.				
5. The vocabulary is appropriate to both the science content and the audience.				

Figure 1.9   An Example of a Performance List Rubric for Oral Presentation for Grades 7-12

Oral Presentation in Science (continued)

Performance Criteria	Assessment			
	Points	Self	Teacher	Other(s)
Optional				
6. Interesting and colorful audiovisual aids or multimedia materials are interwoven to explain and reinforce the screen text and presentation.				
7. The topic is developed completely and thoroughly.				
Presentation				
8. The speaker maintains a proper volume, clear elocution, steady rate, effective inflections and enthusiasm throughout the presentation.				
9. Humor is used positively and in good taste, with consideration given to the composition of the audience.				
10. Stories and motivational scenarios are used appropriately.				
11. Body language such as eye contact, posture, gestures, and body movements are appropriate and are used to create effect.				
12. Delivery is well paced, flows naturally, has good transitions, and is coherent.				
13. The speaker is relaxed, self-confident, and appropriately dressed for purpose or audience.				

(Continued)

Oral Presentation in Science (continued)

Performance Criteria	Assessment			
	Points	**Self**	**Teacher**	**Other(s)**
Audience 14. The audience's attention is maintained by involving them in the presentation.				
15. Information needed by audience to fully understand the presentation is provided.				
16. The speaker gives the audience time to think, reflect, and ask questions about points made in the presentation.				
17. The speaker answers all questions with clear explanations and further elaborations.				
18. The topic and the length of the presentation are appropriate for the audience and within the allotted time limits.				

Comments	Goals	Actions

conceptual understandings through some of their products or performances, such as projects, presentations, or portfolios. And, in all likelihood, you will probably need to reduce your dependence on paper-and-pencil tests. By revising performance tasks over several years, the product only gets sharper and better. This is not necessarily the case with adopting new textbooks every 5 to 10 years.

Teachers often ask in workshops and in college classes, "What is a good source of performance tasks or materials that can be modified to build tasks?" Contemporary science materials, textbooks, and laboratory manuals that emphasize an inquiry-based, constructivist approach, and/or a 5E teaching and learning cycle, are an excellent source of potential performance tasks. Most of the programs endorsed by the National Science Foundation fall into this category. After students have completed an activity, you can very often assess their comprehension of the concept and/or process skills covered by having them repeat the activity with changed variables or hypotheses. Or, you can ask the students to rewrite and improve upon the procedures for completing the activity. Problems or examples from textbooks can be turned into simple performance tasks by asking a student to support his or her answers. Asking questions that probe understanding of why something works in science, and not always how it works, shifts the focus from convergent thinking to divergent thinking and, in so doing, aligns the questions with process goals of instruction.

For example, a group of fifth-grade students is studying concepts of simple machines. By looking at textbook pictures, they can identify various simple machines. They can draw them in their journals—one form of performance. The teacher listens to their conversation and then asks how many simple machines there are within everyday objects, such as skateboards and bicycles. Then, by observing an actual bicycle, the students observe and manipulate its components. Soon, they are able to identify levers and wheels and axles on the bicycle. If the teacher had not asked the question, the students would never have integrated the concepts and skills they were studying.

There are other good sources of information that can be used to construct simple and more complex performance tasks. The National Science Teachers Association publishes many excellent books and journals, such as *Children and Science, Science Scope,* and *Science Teacher,* that are full of good ideas and experiments. Newspapers and magazines often contain articles that can be converted into performance tasks.

Chapter 2

Designing and Using Performance List Rubrics to Assess Student Work

Criteria for Assessment

The assessment tool generally known as the performance list rubric is presented in this chapter as one methodology to assess student achievement on performance tasks. Before developing the performance list rubric, criteria for assessment in general should be considered first. These are as follows:

1. Include only the most important criteria for the performance; do not include trivial, superficial, or superfluous criteria. Avoid excessive length, as an assessment can become dysfunctional through too much detail (Popham, 1997).

2. Be precise and clear in describing the criteria for the performance, and use language that creates a clear mental picture of the performance. Avoid excessively general, amorphous criteria. Do not include statements that are vague, like "fairly complete," "less than a 3," "neat," and so on.

3. Avoid the use of jargon and ambiguous terminology. Use language that is understandable for all audiences.

4. Where possible, illustrate criteria with actual samples of student performance (anchors).

5. Use assessment/evaluation tools that are not task specific (generic) wherever you can (Popham, 1997). Designing an assessment tool for each activity requires much time and energy and may be essentially worthless from an instructional perspective.

6. Avoid the use of quantitative scales for defining criteria. It is difficult to justify that a performance is better because it has, for example, more references. The quality and the relevance of the references are as important as the quantity.

7. Use analytical criteria rather than holistic criteria. Holistic criteria give a single, overall score to the whole performance. Analytical criteria provide separate scores for different dimensions of the performance. Analytical trait systems provide more diagnostic detail, are better suited for monitoring and modifying instruction, and are better understood by students.

8. Develop and use performance list rubrics, for they are the easiest scoring tools to design and are the first step toward designing rubrics later; also, they are the easiest tools for students to use to plan and assess their performance.

When designing assessment tools, a number of important steps need to be addressed before and during the actual design process. These might be summarized as follows.

STEPS FOR DESIGNING ASSESSMENT/EVALUATION TOOLS

1. First, decide whether the activity requires a task-specific or non-task-specific assessment/evaluation tool. If specific factual content (declarative standards and benchmarks) is to be assessed, then, in all

likelihood, you will need a task-specific assessment tool. "Task specific" implies that the tool probably cannot be used for other activities.

2. Next, develop a rough draft of criteria that would define the performance or product.

3. Refine and expand upon the language within the criteria to develop a performance list rubric (could be task specific or non-task specific) for the activity. Define those performance criteria that would reveal student success on the task. Make sure the criteria are clear, essential, teachable, and observable. This list can, and perhaps should, involve students during the design and revision phases. Next, field-test the performance list rubric with students. Look at actual examples of student work to see if you have omitted any important dimensions of the performance. Make necessary revisions.

4. Next, create a holistic rubric using the dimensions from the performance list rubric. The performance list rubric, with some minor modifications, is often used to create the expert level (highest proficiency level) for the performance.

5. Now, create the novice level (lowest proficiency level) of the performance. These two levels of performance define the two extremes of the scoring tool. You may wish to stop here with the holistic rubric until you have actual student work that can be used to create the other levels of performance.

6. Determine which type of rating scale (proficiency level) for the holistic rubric best communicates the judgments you want to make. For example, a numbered scale (such as 4, 3, 2, 1) might be employed. A descriptive scale—such as *extended, satisfactory, partial, minimal, incorrect response, no response*; or *expert, proficient, emergent, novice,* and/or *non-scoreable*—could be used.

7. If it is desirable, now create an analytic rubric. First, cluster the elements of the performance into different concepts. Then, for each conceptual area, create an exemplary (expert) response and an unacceptable (novice) response using the elements from the holistic rubric. Continue with this process until you have completed the analytic rubric.

8. Next, try to score several samples of student work with your draft assessment/evaluation tools.

9. Continue to refine the tools as you collect more student work.

10. Have other teachers use your tools to assess student performance. Continue to refine until all qualities that are important are identified and scored.

11. Share your assessment/evaluation tools with parents and students.

Now, let's examine the format of a performance list rubric; then, we will move on to a discussion of rubrics in Chapter 2. What is a performance list rubric? It is an assessment tool that lists the important criteria of each performance in a checklist template (Educators, 1996). It is analytic in nature in that specific criteria can be assessed as to whether or not they are present in the performance. Single or multiple scores can be given depending upon the desires of the teacher, and it can be generalized or task specific. The table below illustrates some of the advantages and disadvantages of a performance list rubric.

Let's now develop a performance list rubric for a particular task—scientific drawing—a performance that is used extensively in all fields of science to document evidence of student understanding. Scientific drawings are used in life science to record observations such as cell structures, stages of mitosis and meiosis, and results of various investigations. In physics, scientific drawings are used to document concepts such as force and motion, work and machines, and principles of light and sound. In chemistry, scientific drawings are used to depict atomic and molecular structures, chemical reactions and products, and states of matter. In earth/space science, drawings are used to show things such as planetary motion, seasons on Earth, and layers of rock and sedimentation.

Table 2.1   Advantages and Disadvantages of Using a Performance List Rubric

Advantages	Disadvantages
• Provides more diagnostic feedback than holistic tools	• Can be overwhelming when the performance is quite complex—hard to see the forest for the trees
• Provides useful feedback about the strengths and each criterion of the performance	• Does not provide for multiple levels (proficiencies) for weaknesses of each student's performance
• Provides the foundation for creating rubrics (easiest tool for teachers to construct and for students to use)	
• Requires more time to assess performance than a holistic tool	
• Works well for monitoring and adjusting instruction	
• Provides detailed basis for judging performances	
• Works well for self- and peer review	

What are the important dimensions of a scientific drawing? In Step 2 of designing assessment tools, we develop a rough draft of the criteria that would define the performance. Criteria such as the following are all important dimensions of the performance:

- Observation (not inference) and recording of important details
- Accuracy of the scientific concepts shown
- Use of appropriate scale and size
- Labeling of the relevant parts
- Provision of a descriptive title for the drawing
- Use of different perspectives
- Relevant measurements
- Written explanations of what the drawing depicts

Next, we expand these rough criteria into descriptive, unambiguous, and clear statements (Step 3). For example, "observation (not inference) and recording of important details" might be expanded into the written criteria below:

1. The drawing(s) realistically and effectively depict(s) the object(s).

2. The drawing includes only those features that were actually observed and not inferred.

3. Many relevant details are included: size (with metric measurements), colors, textures, shapes, and relationships to surroundings.

"Labeling of the relevant parts" would be expanded to the following:

4. All the parts of the scientific drawing are clearly and accurately labeled.

These statements (criteria) are then displayed and communicated in a template designed for ease of use. Such a template might appear as the performance list rubric that is shown in Figure 2.1.

Scientific Drawing

Name _____ Date _____ Course/Class _____

Task/Assignment _____

Performance Criteria	Assessment			
	Points	Self	Teacher	Other(s)
1. The drawing(s) realistically and effectively depict(s) the object(s).				
2. The drawing includes only those features that were actually observed and not inferred.				
3. Many relevant details are included: size (with metric measurements), colors, textures, shapes, and relationships to surroundings.				
4. Multiple perspectives are drawn to provide the viewer with a complete picture of the structures under study.				
5. A descriptive and accurate title is provided for the drawing(s).				
6. All the parts of the scientific drawing are clearly and accurately labeled.				
7. A detailed, written explanation of what the scientific drawing is intended to show is included.				
8. A key or legend, if needed to explain the drawing(s), is provided.				
9. The scientific drawing(s) is/are of an appropriate size and scale for details to be easily recognized.				
10. A very precise scale and proportion is used consistently. The scale is stated and uses the metric system when possible.				
11. The scientific content is accurately represented and is appropriate for the drawing.				

Comments	Goals	Actions

Figure 2.1   An Example of a Performance List Rubric for a Scientific Drawing

Let's examine several examples of student work and apply the Scientific Drawing Performance List Rubric to assess their efforts.

For Sample 1 (a Grade 7 student), the teacher has chosen to assign 1 point to nine elements of the performance and a 0 to two other elements (#4 and #8). In so doing, the teacher has indicated which elements of this particular performance (Assembling and Drawing a Pendulum System) are important to the performance and are to be assessed, and which elements do not apply or are not being assessed. The nine elements being assessed are weighted equally (1 point each). In Sample 1, the teacher chose not to allow the student to review the Scientific Drawing Performance List Rubric before engaging in the performance. The work of the student is shown in Figure 2.2.

To construct the pendulum, students were given the materials shown in the drawing and were directed to assemble and then draw their set-up of a pendulum system. The student scored his or her work using the performance list rubric, as shown in Figure 2.3. The teacher then collected the student's work and added her assessment, along with several comments (Figure 2.3).

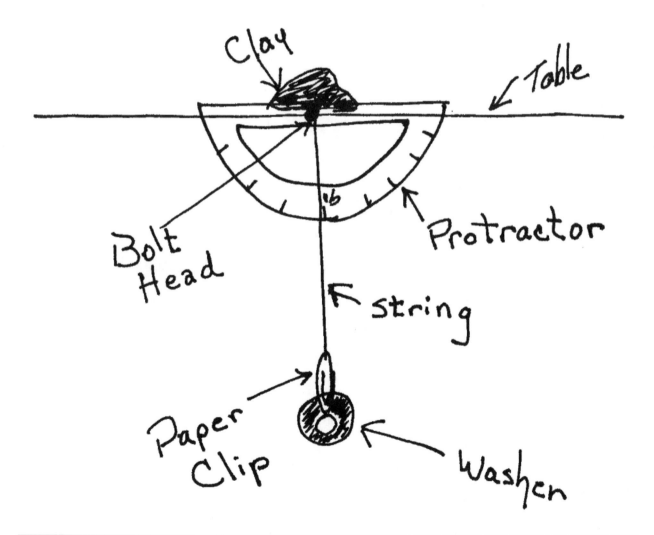

Figure 2.2   Student Sample 1

Scientific Drawing

Name Sample #1 Date ________ Course/Class Grade 7 Integrated Science

Task/Assignment Scientific Drawing of a Pendulum System

Performance Criteria	Assessment			
	Points	Self	Teacher	Other(s)
1. The drawing(s) realistically depict(s) the object(s).	1	1	0	
2. The drawing includes only those features that were actually observed and not inferred.	1	1	0	
3. Many relevant details are included: size (with metric measurements), colors, textures, shapes, and relationships to surroundings.	1	1	0	
4. Multiple perspectives are drawn to provide the viewer with a complete picture of the structures under study.	0			
5. A descriptive and accurate title is provided for the drawing(s).	1	0	0	
6. All the parts of the scientific drawing are clearly and accurately labeled.	1	1	1	

Figure 2.3 Scientific Drawing Performance List for a Pendulum (Sample 1)

Scientific Drawing (continued)

Performance Criteria	Assessment			
	Points	Self	Teacher	Other(s)
7. A detailed, written explanation of what the scientific drawing is intended to show is included.	1	0	0	
8. A key or legend, if needed to explain the drawing(s), is provided.	0			
9. The scientific drawing(s) is/are of an appropriate size and scale for details to be easily recognized.	1	1	1	
10. A very precise scale and proportion is used consistently. The scale is stated and uses the metric system when possible.	1	1	0	
11. The scientific content is accurately represented and is appropriate for the drawing.	1	1	1	

Comments	Goals	Actions

Comments: The drawing does not show all the components of a pendulum system, such as the tabletop that provides the fixed point for the pendulum. No measurements are included in the drawing to give added perspective. Both the title and a written explanation are missing.

Action: Please revise your scientific drawing as part of your homework assignment and then rescore it.

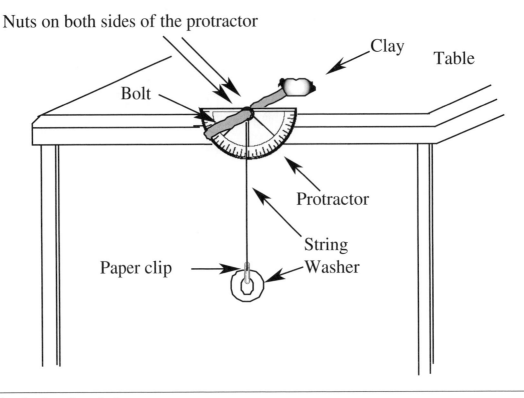

Figure 2.4   Student Sample 2

Shown in Figure 2.4 is an example of another student's work on the same task, in which the performance list rubric had been shared in advance with the student.

For both Student Samples 1 and 2, a non-task-specific performance list was used. However, if the teacher wanted more specific criteria, then a task-specific performance list could be developed. For example, Criterion 1, "The drawing(s) realistically depicts the object(s)," could be revised to "The drawing(s) realistically depict(s) the fixed point for the pendulum, the string, and the mass attached to the end of the pendulum." Criterion 6, "All the parts of the scientific drawing are clearly and accurately labeled," could be revised to "All the parts of the scientific drawing (tabletop, bolt, clay, string, washer, protractor, and paper clip) are clearly and accurately labeled."

WHAT MIGHT A SCIENTIFIC DRAWING PERFORMANCE LIST RUBRIC LOOK LIKE FOR OTHER GRADE LEVELS?

Figures 2.6, 2.7, and 2.8 contain examples of performance list rubrics for Scientific Drawing for other grade levels. Figure 2.6 is a performance list rubric that was developed for Grades K-1. Figure 2.7 is a performance list rubric for Grades 2-3, and Figure 2.8 is for Grades 4-6. The same conceptual criteria were used for planning and developing each performance list rubric, but the final statements are written at increasing levels of sophistication.

As presented in this chapter, performance list rubrics provide clear criteria for assessing performance and are a necessary first step for the construction of other assessment tools, such as rubrics. Performance list rubrics can be used for self-, teacher, or other (peer, scientist, etc.) assessments.

Often, it is desirable to use assessment tools in other formats, such as holistic or analytic rubrics. A discussion of holistic and analytic rubrics, their construction from performance list rubrics, and their use follows in Chapter 3.

Scientific Drawing

Name ____Sample #2____   Date _____   Course/Class Grade 7 Integrated Science

Task/Assignment ___Scientific Drawing of a Pendulum System___

Performance Criteria	Assessment			
	Points	Self	Teacher	Other(s)
1. The drawing(s) realistically depict(s) the object(s).	1	1	1	
2. The drawing includes only those features that were actually observed and not inferred.	1	1	1	
3. Many relevant details are included: size (with metric measurements), colors, textures, shapes, and relationships to surroundings.	1	1	1	
4. Multiple perspectives are drawn to provide the viewer with a complete picture of the structures under study.	0			
5. A descriptive and accurate title is provided for the drawing(s).	1	1	1	
6. All the parts of the scientific drawing are clearly and accurately labeled.	1	1	1	

Figure 2.5   Scientific Drawing Performance List for a Pendulum (Sample 2)

(Continued)

Scientific Drawing (continued)

Performance Criteria	Assessment			
	Points	Self	Teacher	Other(s)
7. A detailed, written explanation of what the scientific drawing is intended to show is included.	1	1	1	
8. A key or legend, if needed to explain the drawing(s), is provided.	0			
9. The scientific drawing(s) is/are of an appropriate size and scale for details to be easily recognized.	1	1	1	
10. A very precise scale and proportion is used consistently. The scale is stated and uses the metric system when possible.	1	1	1	
11. The scientific content is accurately represented and is appropriate for the drawing.	1	1	1	

Comments	Goals	Actions
Comments: The drawing is an excellent representation of the pendulum system. Good work! No additional work or revisions are needed.		

Name _____

Date _____

Drawing of _____

Scientific Drawing

	Student			Teacher		
1. Did I draw everything I saw?	Great	O. K.	Needs Work	Great	O. K.	Needs Work
2. Did I label all the parts of the drawing?	Great	O. K.	Needs Work	Great	O. K.	Needs Work
3. Did I give the drawing a title?	Great	O. K.	Needs Work	Great	O. K.	Needs Work
4. Did I make the drawing large enough to see all the parts clearly?	Great	O. K.	Needs Work	Great	O. K.	Needs Work
5. Is my drawing neat and easy to follow?	Great	O. K.	Needs Work	Great	O. K.	Needs Work

Teacher Comments:

Figure 2.6   Performance List Rubric for Scientific Drawing for Grades K–1

Scientific Drawing

Name _____

Date _____

Topic _____

	Performance Criteria	Assessment		
		Points	Self	Teacher
1.	My scientific drawing shows details of what I actually observed.			
2.	All parts of my scientific drawing are clearly and accurately labeled.			
3.	My drawing has a title that explains what the drawing is all about.			
4.	My scientific drawing is large enough to see all parts clearly.			

Teacher Comments:

Figure 2.7   Performance List Rubric for Scientific Drawing for Grades 2-3

Scientific Drawing

Name _____ Date _____ Course/Class _____

Task/Assignment _____

Performance Criteria	Assessment			
	Points	Self	Teacher	Other(s)
1. My scientific drawing looks similar to what I observed.				
2. I included as many details as possible: colors, textures, shapes, measurements, etc.				
3. I labeled all the parts of my scientific drawing.				
4. I wrote a title that tells what my scientific drawing shows.				
5. I provided a written explanation of what my scientific drawing is intended to show.				
6. My scientific drawing is of an appropriate size for details to be easily recognized.				

Comments	Goals	Actions

Figure 2.8   Performance List Rubric for Scientific Drawing for Grades 4-6

Chapter 3

Designing and Using Scoring Rubrics to Assess Student Work

SCORING RUBRICS

There is some disagreement about the origins of the word *rubric.* Some contend that the word has its origins in the Latin word *rubrica,* which refers to the use of "red earth" to mark things of significance (Marzano, Pickering, & McTighe, 1993). Others contend that *rubric* has its origins in the work of Christian monks who reproduced sacred literature and marked each major section of a copied book with a large red letter. Regardless of its origins, all agree that the original meaning of rubric has little to do with assessment of student work. Today, the word *rubric* generally refers to a fixed scale where each point on the scale clearly describes the quality of a product or performance. Rubrics are frequently accompanied by examples (anchors) of products or performances to illustrate the various points on the scale.

According to Popham (1997), three essential elements form the skeleton of a rubric: evaluative criteria, quality definitions, and a scoring strategy. These are used to develop the rubric, regardless of whether the scoring strategy is holistic or analytic.

Evaluative criteria, as outlined in Chapter 2, are used to distinguish expert work from unacceptable work. Using the example of the Scientific Drawing performance list rubric from Chapter 2, evaluative criteria would include factors such as observation (not inference) and recording of important details, accuracy of the scientific concepts shown, use of appropriate scale and size, labeling of the relevant parts, provision of a descriptive title for the drawing, use of different perspectives, relevant measurements, and written explanations of what the drawing depicts.

Quality definitions are used to establish the different levels of performance. To earn the maximum number of points, all the parts of the scientific drawing are clearly and accurately labeled. A scoring rubric then must provide a separate description for each level of performance. These levels are assigned either numbers (4, 3, 2, 1) or descriptive terms such as *expert, proficient, emergent,* and *novice.*

A scoring strategy may either be holistic or analytic, depending upon its purposes and the nature of the feedback to be provided to the students and teacher. In holistic scoring, one single score, which consists of an aggregate of all the evaluative criteria, is provided. Analytic scoring provides multiple scores, depending upon the number of evaluative criteria.

Consequently, scoring rubrics typically have been classified as either holistic or analytic, with each having a unique format. However, regardless of the type, both are easier to design once evaluative criteria have been developed in the form of a performance list or performance list rubric, as discussed in Chapter 2.

CHARACTERISTICS OF HOLISTIC RUBRICS

- Describe the student's work taken as a whole—an overall general impression
- Can be generalized or task specific
- Capture the essence (gist) of the performance

Table 3.1   Advantages and Disadvantages of Holistic Rubrics

Advantages	Disadvantages
• Faster and easier to develop and use than analytic rubrics	• Can be hard to use unless you've developed it yourself
• More intuitive – can sort work into groups or piles based on perceived quality	• Can be perceived as more subjective, so it is more open to challenge
• Good for complex performances	• Does not give students specific feedback on what to improve
• Quick scoring	

• Provide a single score based on the overall quality of the work
• Are most effective when used with anchors or exemplars of student performances

Table 3.1 illustrates the strengths and weaknesses of this form of rubric.

An example of a holistic rubric, developed from the Scientific Drawing performance list from Chapter 2, is shown in Figure 3.1. Criteria from the performance list rubric were summarized and consolidated to form a succinct narrative for the highest level of the holistic rubric. In this case, descriptive terms, accompanied by a numeric value, were used to describe the different proficiency levels of the performance—*expert* (4), *proficient* (3), *emergent* (2), and *novice* (1).

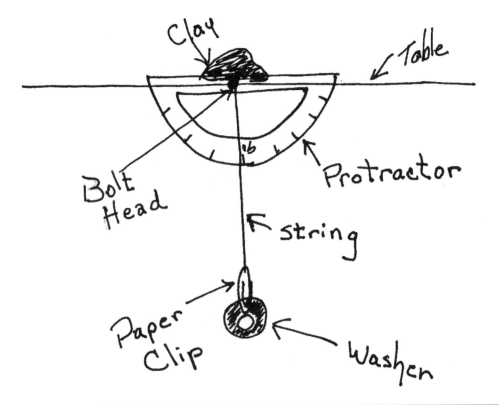

Student Sample 1

Scientific Drawing

Name _____ Date _____ Course/Class _____

Task/Assignment _____

Expert 4	The drawing(s) realistically and effectively depict(s) the object(s). Multiple perspectives are provided to enhance understanding. A descriptive and accurate title is provided and all the parts of the drawing are clearly labeled. A detailed written explanation of what the scientific drawing is intended to show is included, along with a key or legend to further explain the drawing(s). The drawing(s) is/are of an appropriate size and consistent metric scale for details to be easily recognized. The scientific content is accurately represented and is appropriate for the drawing.
Proficient 3	The drawing(s) depict(s) the object(s). Many details are included. A descriptive and accurate title is provided and most parts of the drawing are clearly and neatly labeled. A sketchy written explanation of what the scientific drawing is intended to show is included. The drawing(s) is/are of an appropriate size and scale for details to be easily recognized. The scientific content is accurately represented and is appropriate for the drawing.
Emergent 2	The drawing(s) reasonably depict(s) the object(s). The drawing(s) is/are a reasonable rendition of the object(s), but may include features that were not actually observed. Some details are included. Only one perspective of the object(s) is provided. A title is provided for the drawing(s). Some parts of the scientific drawing are labeled. Labeling lacks neatness, legibility, and attractiveness. A sketchy written explanation of what the scientific drawing is intended to show is included. The drawing(s) is/are inappropriately sized and scaled. The scientific content contains some inaccuracies.
Novice 1	The drawings are clearly lacking in realism, accuracy, and detail. It is difficult to tell what the drawing(s) represent(s). Scale and proportion are clearly lacking. Metric measurements are missing. Few distinguishing forms, structures, and details are labeled. Labeling is not consistently neat, legible, and attractive. No attempt is made to provide a title of the drawing(s). The scientific content contains many inaccuracies.

Comments	Goals	Actions

Figure 3.1   Holistic Rubric for Scientific Drawing

Using Student Sample 1 from the first chapter (shown below) and applying the holistic rubric for Scientific Drawing, the student would have received a novice (1) rating. The characteristics of the drawing closely parallel that of the novice level in that no title is included, no description of what the drawing depicts is included, no measurements are stated, and the labeling of the parts of the pendulum is not consistently neat and legible.

Now, let's turn our attention to another type of rubric, known as the analytic rubric. How is it distinguished from the holistic rubric?

CHARACTERISTICS OF ANALYTIC RUBRICS

- Describe the different dimensions of the performance
- Provide separate ratings for each of the different dimensions of the performance (can be generalized or task specific)
- Provide multiple scores on the performance

Table 3.2 shows the strengths and weaknesses of analytic rubrics.

An example of an analytic rubric constructed from the traits of the holistic rubric of Scientific Drawing is shown in Figure 3.2. The small boxes within each cell are used for self-assessment, peer assessment, and/or teacher assessment. Using Sample 1 again and applying the analytic rubric, the student would have received the following ratings:

- Emergent (2) for Accuracy and Realism
- Novice (1) for Scale and Proportion
- Emergent (2) for Labeling
- Novice (1) for Titles and Accompanying Text

As can be evidenced from scoring the same sample with both holistic and analytic rubrics, somewhat different scores can emerge. This is a reflection of the design of the two different tools. The holistic tool, although quicker to use, does not provide the diagnostic details to let students know why they received a "1" instead of a "2." However, if the teacher adds comments or designates goals or actions to be taken by the student, then this shortcoming of the holistic tool can be somewhat ameliorated.

Chapter 4 contains actual examples of performance list rubrics, holistic rubrics, and analytic rubrics for many performances that are common in performance-based science classes. They have been field-tested and revised for several years with students in Grades K-12.

Table 3.2   Advantages and Disadvantages of Analytic Rubrics

Advantages	Disadvantages
- Provides useful feedback about the strengths and weaknesses of each student's performance and of the instruction (more diagnostic)	- Can be time-consuming to construct and use - Can make it hard to see the forest for the trees
- Provides detailed basis for judging performances	
- Provides extra details when multiple scorers are scoring the same piece of work	
- Provides extra details for multiple grade levels to emphasize the same criteria	

Scientific Drawing (Analytic Rubric)

Name _____ Date _____ Course/Class _____

Task/Assignment _____

	Accuracy and Realism	Scale and Proportion	Labeling	Titles and Accompanying Text
Expert **(4)**	The drawing(s) realistically and effectively depict(s) the object(s). Amazing detail is provided for size, color, texture, and shape. Multiple perspectives are provided to clearly distinguish form, structures, and dimensions. The scientific content is accurately represented and is appropriate for the drawing. ☐	A very precise scale and proportion is provided using metric measurements. The scale and proportion are appropriate for showing details. The scale is stated either in the drawing itself or the accompanying key or legend. The relationship between the object and its environment is shown. ☐	All distinguishing forms, structures, and details are clearly labeled. Labeling is consistently neat, legible, and attractive in appearance. ☐	A descriptive and accurate title of the drawing(s) is provided. A detailed, interpretative, written explanation of what the drawing(s) is/are intended to show is provided. ☐
Proficient **(3)**	The drawing(s) depict(s) the object(s). Amazing detail is provided for size, color, texture, and shape. Multiple perspectives are missing. The scientific content is accurately represented and is appropriate for the drawing. ☐	A very precise scale and proportion is provided using metric measurements. The scale and proportion are appropriate for showing details. The legend or key of the scale is missing. The relationship between the object and its environment is shown. ☐	All distinguishing forms, structures, and details are clearly labeled. Labeling is not consistently neat, legible, and attractive in appearance. ☐	A title of the drawing(s) is provided. A written explanation of what the drawing(s) is/are intended to show is provided. However, in both cases, details and clarity are lacking. ☐

Figure 3.2 Analytic Rubric for Scientific Drawing

(Continued)

Scientific Drawing (continued)

	Accuracy and Realism	Scale and Proportion	Labeling	Titles and Accompanying Text
Emergent (2)	The drawing(s) reasonably depict(s) the object(s). Many details are provided. Multiple perspectives are missing. The scientific content contains some inaccuracies. ❑	A rather imprecise scale and proportion is provided. Metric measurements are missing. The scale and proportion are appropriate for showing details. The legend or key of the scale is missing. The relationship between the object and its environment is not shown. ❑	Most distinguishing forms, structures, and details are clearly labeled. Labeling is not consistently neat, legible, and attractive in appearance. ❑	An attempt is made to provide a title of the drawing(s). A written explanation of what the drawing(s) is/are intended to show is not provided. ❑
Novice (1)	The drawings are clearly lacking in how realistic the object(s) is/are drawn. It is difficult to tell from the drawing(s) what the object is. Few details are provided. Multiple perspectives are missing. The scientific content contains many inaccuracies. ❑	Scale and proportion are clearly lacking. Metric measurements are missing. The scale and proportion are not appropriate for showing details. The legend or key of the scale is missing. The relationship between the object and its environment is not shown. ❑	Few distinguishing forms, structures, and details are clearly labeled. Labeling is not consistently neat, legible, and attractive in appearance. ❑	No attempt is made to provide a title of the drawing(s). A written explanation of what the drawing(s) is/are intended to show is not provided. ❑

Chapter 4

Assessment Tools for Science Grades K-1

Name _____

Date _____

Collecting Data on _____

Collecting Data

	Student			Teacher		
1. Did I record my data correctly?	Great	O. K.	Needs Work	Great	O. K.	Needs Work
2. Did I use the correct units when I recorded my data?	Great	O. K.	Needs Work	Great	O. K.	Needs Work
3. Did I collect my data carefully?	Great	O. K.	Needs Work	Great	O. K.	Needs Work
4. Did I make my data collection clear and complete?	Great	O. K.	Needs Work	Great	O. K.	Needs Work

Teacher Comments:

Figure 4.1   Collecting Data

Name _____

Date _____

Cooperative Learning on _____

Cooperative Learning

	Student	Teacher
1. Did I do my job in my group?	Great O. K. Needs Work	Great O. K. Needs Work
2. Did I follow directions?	Great O. K. Needs Work	Great O. K. Needs Work
3. Did I finish my part on time?	Great O. K. Needs Work	Great O. K. Needs Work
4. Did I help others in my group?	Great O. K. Needs Work	Great O. K. Needs Work

Figure 4.2   Cooperative Learning

Cooperative Learning (continued)

	Student	Teacher
5. Did I listen to others in my group?	Great O. K. Needs Work	Great O. K. Needs Work
6. Did I get along with others in my group?	Great O. K. Needs Work	Great O. K. Needs Work
7. Did I help my group clean up?	Great O. K. Needs Work	Great O. K. Needs Work

Teacher Comments:

Name _____

Date _____

Concept Map on _____

Concept Map

		Student			Teacher	
1.	Did my concept map go from the big idea to the little ideas?	Great	O. K.	Needs Work	Great	O. K.	Needs Work
2.	Did I use linking words that make sense and connect the ideas?	Great	O. K.	Needs Work	Great	O. K.	Needs Work
3.	Did my concept map have a title?	Great	O. K.	Needs Work	Great	O. K.	Needs Work
4.	Did I make my concept map neat?	Great	O. K.	Needs Work	Great	O. K.	Needs Work

Teacher Comments:

Figure 4.3   Concept Map

Name _____

Date _____

Data Collection Table on _____

Data Collection Table

	Student			Teacher		
1. Did I give my data collection table a title?	Great	O. K.	Needs Work	Great	O. K.	Needs Work
2. Did I label all the headings?	Great	O. K.	Needs Work	Great	O. K.	Needs Work
3. Did I put my data under the correct heading?	Great	O. K.	Needs Work	Great	O. K.	Needs Work
4. Did I label all data with the correct units?	Great	O. K.	Needs Work	Great	O. K.	Needs Work
5. Did I make the data collection table easy to read?	Great	O. K.	Needs Work	Great	O. K.	Needs Work

Teacher Comments:

Figure 4.4   Data Collection Table

Name _____

Date _____

Drawing Conclusions on _____

Drawing Conclusions

	Student	Teacher
1. Did I answer the original question?	Great O. K. Needs Work	Great O. K. Needs Work
2. Did I use data from my data collection table to support my conclusion?	Great O. K. Needs Work	Great O. K. Needs Work
3. Did I write my conclusion neatly?	Great O. K. Needs Work	Great O. K. Needs Work

Teacher Comments:

Figure 4.5   Drawing Conclusions

Name _____

Date _____

Graph of _____

Graphing

	Student			Teacher		
1. Did I give the graph a title?	Great	O. K.	Needs Work	Great	O. K.	Needs Work
2. Did I label all the parts of the graph?	Great	O. K.	Needs Work	Great	O. K.	Needs Work
3. Did I put my data in the correct place on the graph?	Great	O. K.	Needs Work	Great	O. K.	Needs Work
4. Did I make the graph neat and easy to read?	Great	O. K.	Needs Work	Great	O. K.	Needs Work

Teacher Comments:

Figure 4.6   Graphing

Name _____

Date _____

Making a Model of _____

Making a Model

	Student			Teacher		
1. Did my model show the correct idea?	Great	O. K.	Needs Work	Great	O. K.	Needs Work
2. Did I tell what my model shows?	Great	O. K.	Needs Work	Great	O. K.	Needs Work
3. Did I make my model easy to understand?	Great	O. K.	Needs Work	Great	O. K.	Needs Work
4. Did I make my model neat?	Great	O. K.	Needs Work	Great	O. K.	Needs Work

<u>**Teacher Comments:**</u>

Figure 4.7 Making a Model

Name _____

Date _____

Observing _____

Observing

	Student	Teacher
1. Did I use my senses (see, hear, touch, smell) to help me?	Great O. K. Needs Work	Great O. K. Needs Work
2. Did I write exactly what I saw?	Great O. K. Needs Work	Great O. K. Needs Work
3. Did I use the right equipment to help me?	Great O. K. Needs Work	Great O. K. Needs Work
4. Did I follow all the safety rules?	Great O. K. Needs Work	Great O. K. Needs Work

<u>Teacher Comments</u>:

Figure 4.8   Observing

Name _____

Date _____

Predicting about _____

Predicting

	Student			Teacher		
1. Did my prediction answer the question?	Great	O. K.	Needs Work	Great	O. K.	Needs Work
2. Did I explain my prediction using the word "because"?	Great	O. K.	Needs Work	Great	O. K.	Needs Work
3. Is my explanation based on what I know or have learned?	Great	O. K.	Needs Work	Great	O. K.	Needs Work
4. Did I start my prediction with "I think" or "I believe"?	Great	O. K.	Needs Work	Great	O. K.	Needs Work
5. Did I start my prediction with a capital letter and end it with a period?	Great	O. K.	Needs Work	Great	O. K.	Needs Work

Teacher Comments:

Figure 4.9   Predicting

Name _____

Date _____

Drawing of: _____

Scientific Drawing

	Student			Teacher		
1. Did I draw everything I saw?	Great	O. K.	Needs Work	Great	O. K.	Needs Work
2. Did I label all the parts of the drawing?	Great	O. K.	Needs Work	Great	O. K.	Needs Work
3. Did I give the drawing a title?	Great	O. K.	Needs Work	Great	O. K.	Needs Work
4. Did I make the drawing large enough to see all the parts clearly?	Great	O. K.	Needs Work	Great	O. K.	Needs Work
5. Is my drawing neat and easy to follow?	Great	O. K.	Needs Work	Great	O. K.	Needs Work

__Teacher Comments:__

Figure 4.10   Scientific Drawing

Chapter 5

Assessment Tools for Science Grades 2-3

Collecting Scientific Data

Name _____

Date _____

Topic _____

	Performance Criteria	Assessment		
		Points	Self	Teacher
1.	I used a data collection table to organize my data.			
2.	I completed all parts of my data collection table.			
3.	I made my measurements carefully.			
4.	I recorded my measurements using the correct units.			
5.	I did the same experiment at least 3 times to collect my scientific data.			

Teacher Comments:

Figure 5.1 Collecting Scientific Data

Cooperative Learning in Science

Name _____

Date _____

Topic _____

	Performance Criteria	Assessment		
		Points	Self	Teacher
1.	I did my job.			
2.	I got along with others in my group.			
3.	I followed my teacher's directions.			
4.	I listened to others and their ideas.			
5.	My group finished on time.			

Teacher Comments:

Figure 5.2   Cooperative Learning in Science

Concept Map

Name _____

Date _____

Topic _____

	Performance Criteria	Assessment		
		Points	Self	Teacher
1.	My choice of words is related to the science concept.			
2.	I included appropriate linking words that connect the concepts.			
3.	My concept map flows from the big idea to little ideas.			
4.	My concept map has a title that tells about the big idea.			

Teacher Comments:

Figure 5.3   Concept Map

Data Collection Table

Name _____

Date _____

Topic _____

	Performance Criteria	Assessment		
		Points	Self	Teacher
1.	I gave my data collection table a title that tells about what the table displays.			
2.	All the heading/rows/columns of data are labeled.			
3.	The collected data were recorded under the correct heading/row/on column.			
4.	All measurements are labeled with the correct unit.			
5.	The data table is neat, clear, and easy to read.			

Teacher Comments:

Figure 5.4   Data Collection Table

Graphing Scientific Data

Name _____

Date _____

Topic _____

	Performance Criteria	Assessment		
		Points	**Self**	**Teacher**
1.	I gave my graph a title that tells what the data show.			
2.	I labeled all the parts of my graph (units of measurement, rows, columns, etc.).			
3.	My data are correctly placed on the graph.			
4.	My graph is easy to read.			

Teacher Comments:

Figure 5.5   Graphing Scientific Data

Journal Entry

Name _____

Date _____

Topic _____

	Performance Criteria	Assessment		
		Points	Self	Teacher
1.	My journal entry has a heading including date, name, and class.			
2.	My journal entry has at least 3 sentences or more that tell about what I learned today.			
3.	My journal entry is clear and complete.			

<u>**Teacher Comments**</u>:

Figure 5.6   Journal Entry

Observing in Science (Performance List Rubric)

Name _____

Date _____

Topic _____

Performance Criteria	Assessment		
	Points	**Self**	**Teacher**
1. I used my senses (see, hear, touch, smell) to help me. (Taste was used only with the permission of the teacher.)			
2. I wrote exactly what I observed.			
3. I used the proper equipment to help me.			
4. I followed all safety rules when making my scientific observations.			

Teacher Comments:

Figure 5.7   Observing in Science (Performance List Rubric)

Name _____

Date _____

Topic _____

Observing in Science (Analytic Rubric)

	Senses	Writing	Equipment	Safety
3	I used appropriate senses to help me. ☐	I wrote down everything exactly as I saw it. ☐	I used proper equipment to help me. ☐	I followed all safety rules when making my scientific observations. ☐
2	I used some of the appropriate senses to help me. ☐	I wrote down only some of what I saw. ☐	I mostly used proper equipment to help me. ☐	I followed most of the safety rules when making my scientific observations. ☐
1	I did not use appropriate senses to help me ☐	I wrote down a little of what I saw or nothing at all. ☐	I did not use proper equipment to help me. ☐	I did not follow all safety rules when making my scientific observations. ☐

Teacher Comments:

Figure 5.8   Observing in Science (Analytic Rubric)

Predicting in Science

Name _____

Date _____

Topic _____

	Performance Criteria	Assessment		
		Points	Self	Teacher
1.	My prediction is related to my question.			
2.	My prediction is based on research on what I already know or have seen before.			
3.	I started my prediction with **I think** or **I believe.**			
4.	My prediction is written as a clear and complete sentence that begins with a capital letter and ends with a period.			

<u>**Teacher Comments**</u>:

Figure 5.9   Predicting in Science

Scientific Drawing

Name _____

Date _____

Topic _____

	Performance Criteria	Assessment		
		Points	**Self**	**Teacher**
1.	My scientific drawing shows details of what I actually observed.			
2.	All parts of my scientific drawing are clearly and accurately labeled.			
3.	My drawing has a title that explains what the drawing is all about.			
4.	My scientific drawing is large enough to see all parts clearly.			

Teacher Comments:

Figure 5.10   Scientific Drawing

Written Summary of a Graph

Name _____

Date _____

Topic _____

	Performance Criteria	Assessment		
		Points	**Self**	**Teacher**
1.	My summary tells what the graph is about.			
2.	My summary gives at least 2 important details about the graph.			
3.	My summary is only about the information presented on the graph.			
4.	I used complete sentences when writing my summary.			
5.	My summary is clear and complete.			

Teacher Comments:

Figure 5.11   Written Summary of a Graph

Chapter 6

Assessment Tools for Science Grades 4-6

Collecting and Organizing Data

Name _____ Date _____ Course/Class _____

Task/Assignment _____

Performance Criteria	Assessment			
	Points	Self	Teacher	Other(s)
1. Measurements were carefully made so as to increase accuracy.				
2. The set of data is recorded in a list, table, or chart.				
3. The data are recorded neatly, accurately, and completely in a list, table, or chart.				
4. The set of data shows that there were repeated trials of the experiment.				
5. All measurements are labeled with both the quantity and an appropriate unit.				
6. The mean (average) was calculated accurately for the repeated trials.				

Comments	Goals	Actions

Figure 6.1   Collecting and Organizing Data

Concept Map

Name _____   Date _____   Course/Class _____

Task/Assignment _____

Performance Criteria	Assessment			
	Points	Self	Teacher	Other(s)
1. The most general concept is at the top of the map.				
2. I included appropriate linking words to connect the concepts.				
3. The linking words used to connect concept words make the connection between the two concept words understandable.				
4. My concept map has an appropriate title that tells what the specific main idea is on the map.				
5. My concept map is easy to follow.				

Comments	Goals	Actions

Figure 6.2   Concept Map

Cooperative Learning

Name _____ Date _____ Course/Class _____

Task/Assignment _____

Performance Criteria	Assessment			
	Points	Self	Teacher	Other(s)
1. I did my job.				
2. I got along with others in my group.				
3. My group received at least one compliment or praise.				
4. I listened to others and their ideas.				
5. My group finished all that was expected.				
6. I did better in the group than if I had worked alone.				

Comments	Goals	Actions

Figure 6.3 Cooperative Learning

Data Analysis

Name _____ Date _____ Course/Class _____

Task/Assignment _____

Performance Criteria	Assessment			
	Points	**Self**	**Teacher**	**Other(s)**
1. The analysis includes all the important data elements.				
2. Appropriate mathematical procedures were selected and applied.				
3. All mathematical calculations are shown and are accurate.				
4. Appropriate graphic display techniques were used to further analyze the data.				
5. The analysis revealed any significant patterns in the data.				
6. Appropriate inferences and/or conclusions were made based on the data analysis.				

Comments	Goals	Actions

Figure 6.4   Data Analysis

Data Collection Table

Name _____ Date _____ Course/Class _____

Task/Assignment _____

Performance Criteria	Assessment			
	Points	Self	Teacher	Other(s)
1. The data collection table displays all the **collected** data.				
2. The data collection table has a title that reflects what the table displays.				
3. All the heading/rows/columns of data are correctly labeled.				
4. The set of data was recorded under the appropriate heading/row/column.				
5. All measurements are labeled with the correct unit.				
6. The data table is clear and complete.				

Comments	Goals	Actions

Figure 6.5 Data Collection Table

Designing a Scientific Experiment

Name _____ Date _____ Course/Class _____

Task/Assignment _____

Performance Criteria	Assessment			
	Points	**Self**	**Teacher**	**Other(s)**
1. There is a testable question for the experiment.				
2. Research (literature review) was done to learn more about the question.				
3. The design of the experiment tests the hypothesis.				
4. A list of all necessary materials is included.				
5. A detailed step-by-step procedure is included.				
6. The procedures are written clearly enough so that another person could repeat the experiment.				
7. The procedure shows that repeated trials were done.				
8. Data were collected and recorded for each trial.				
9. An appropriate graph was created to display the data.				
10. Conclusions were drawn using the data.				
11. The conclusions refer back to the hypothesis.				
12. Ideas for future research are included.				
13. A log or journal was used to record observations.				
14. A three- or more sentence summary was written explaining and describing what was discovered or learned.				

Comments	Goals	Actions

Figure 6.6   Designing a Scientific Experiment

Graphing Scientific Data

Name _____ Date _____ Course/Class _____

Task/Assignment _____

Performance Criteria	Assessment			
	Points	Self	Teacher	Other(s)
1. I used an appropriate type of graph (bar graph, pictograph, stem-and-leaf, circle graph, line plot graph, etc.).				
2. The title of my graph clearly relates to the information displayed on the graph.				
3. I used my data to choose an appropriate interval to number my x axis and y axis (2's, 3's, 5's, 10's, 100's, etc.).				
4. When placing the numbers on my graph, I spaced them evenly.				
5. I labeled all the parts of my graph (units of measurement, x and y axis, columns, rows, etc.).				
6. My set of data is plotted on the graph accurately.				
7. My graph is clear and complete.				

Comments	Goals	Actions

Figure 6.7 Graphing Scientific Data

Hypothesizing

Name _____ Date _____ Course/Class _____

Task/Assignment _____

Performance Criteria	Assessment			
	Points	**Self**	**Teacher**	**Other(s)**
1. My hypothesis is directly related to the question.				
2. My hypothesis is a simple statement that is based on research and/or what I already know about the question.				
3. My hypothesis states what I believe will happen and why.				
4. My hypothesis is a clear declarative statement.				
5. My hypothesis is written as a complete sentence beginning with a capital letter and ending with a period.				

Comments	Goals	Actions

Figure 6.8   Hypothesizing

Observing and Drawing Conclusions

Name _____   Date _____   Course/Class _____

Task/Assignment _____

Performance Criteria	Assessment			
	Points	Self	Teacher	Other(s)
1. All appropriate senses (**except** taste) were used to make observations.				
2. Appropriate scientific tools and materials were used to make the observations.				
3. Correct metric measurements were taken when necessary.				
4. Both the quantity and the unit for each measurement were recorded.				
5. Observations were based on what was actually observed and not inferred.				
6. Collected data were recorded and organized clearly and accurately.				
7. Reasonable conclusions were drawn using observations, collected data, and what was already known.				

Comments	Goals	Actions

Figure 6.9   Observing and Drawing Conclusions

PowerPoint Presentation

Name _____ Date _____ Course/Class _____

Task/Assignment _____

Performance Criteria	Assessment			
	Points	**Self**	**Teacher**	**Other(s)**
1. The topic has been extensively and accurately researched.				
2. A storyboard, consisting of logically and sequentially numbered slides, has been developed.				
3. The introduction is interesting and engages the audience.				
4. The fonts are easy to read and point size varies appropriately for headings and text.				
5. The use of italics, bold, and underline contributes to the readability of the text.				
6. The background and colors enhance the text.				
7. The graphics, animations, and sounds enhance the overall presentation.				
8. Graphics are of proper size.				
9. The text is free of spelling, punctuation, capitalization, and grammatical errors.				

Comments	Goals	Actions

Figure 6.10   PowerPoint Presentation

Science Fair Display

Name _____ Date _____ Course/Class _____

Task/Assignment _____

Performance Criteria	Assessment			
	Points	**Self**	**Teacher**	**Other(s)**
1. Overall appearance is neat and attractive.				
2. All necessary parts are labeled (Question, Hypothesis, Materials, Procedure, Data, Summary of Results, and Conclusion).				
3. I used no more than 3 colors when doing my backboard.				
4. My backboard has a title.				
5. I remembered to write or attach the information about myself on the back of my backboard.				
6. All of the words on my backboard are spelled correctly.				
7. All necessary parts are included on my graph.				

Figure 6.11   Science Fair Display

(Continued)

Science Fair Display (continued)

Performance Criteria	Assessment			
	Points	Self	Teacher	Other(s)
8. I included a written summary of the data I collected.				
9. The conclusion reflects back to the hypothesis and states whether the hypothesis was correct or incorrect and why.				
10. My conclusion only states the one major finding of the experiment.				
11. I listed at least one question for further study in my conclusion.				
12. The research paper on my topic is placed in front of my backboard.				
13. My journal/log containing notes, observations, and data collected during my experiment is also placed in front of the backboard.				

Comments	Goals	Actions

Scientific Drawing

Name _____ Date _____ Course/Class _____

Task/Assignment _____

Performance Criteria	Assessment			
	Points	Self	Teacher	Other(s)
1. My scientific drawing looks similar to what I observed.				
2. I included as many details as possible: colors, textures, shapes, measurements, etc.				
3. I labeled all the parts of my scientific drawing.				
4. I wrote a title that tells what my scientific drawing shows.				
5. I provided a written explanation of what my scientific drawing is intended to show.				
6 My scientific drawing is of an appropriate size for details to be easily recognized.				

Comments	Goals	Actions

Figure 6.12   Scientific Drawing

Web Site Design and Use

Name _____ Date _____ Course/Class _____

Task/Assignment _____

Performance Criteria	Assessment			
	Points	Self	Teacher	Other(s)
1. The purpose of the site is quite obvious.				
2. The site creates an excellent first impression.				
3. The layout is easy to navigate and is user friendly.				
4. Useful content is not more than three clicks away from the home page.				
5. Graphics and color add to the instructional value.				
6. The font size varies appropriately for headings and text. The use of italics, bold, and underline contributes to the readability of the text.				
7. The text is easy to read and the language used is appropriate for the intended audience.				
8. Content is accurate, complete, and useful.				
Optional The site invites feedback and provides an e-mail address for the contact person.				
The site announces the last time it was updated and links have been kept current.				

Figure 6.13 Web Site Design and Use

Writing Procedures for Scientific Investigations

Name _____ Date _____ Course/Class _____

Task/Assignment _____

Performance Criteria	Assessment			
	Points	Self	Teacher	Other(s)
1. The procedures are written in a step-by-step, easy-to-follow format so that another person could repeat this experiment.				
2. Scientific drawings or other graphics are added to help explain the procedures.				
3. A list of necessary materials and equipment is included.				
4. Any safety issues or precautions are listed.				
5. Dependent, independent, and control variables have been identified within the steps.				
6. Only one independent variable is tested at a time.				
7. The steps show that repeated trials were done.				
8. Language is clear, grammatically correct, and written in complete sentences.				

Comments	Goals	Actions

Figure 6.14   Writing Procedures for Scientific Investigations

Written Summary of a Graph

Name _____   Date _____   Course/Class _____

Task/Assignment _____

Performance Criteria	Assessment			
	Points	Self	Teacher	Other(s)
1. I stated the topic of the graph.				
2. I wrote at least 3 sentences describing what the graph displays.				
3. A clear explanation of any outliers is included.				
4. My summary is only about the information presented on the graph.				
5. I used clear and complete sentences when writing my summary.				

Comments	Goals	Actions

Figure 6.15   Written Summary of a Graph

Chapter 7

Assessment Tools for Science Grades 7-12

Brief Constructed Response to Open-Ended Science Questions (Performance List Rubric)

Name _____   Date _____   Course/Class _____

Task/Assignment _____

Performance Criteria	Assessment			
	Points	Self	Teacher	Other(s)
1. The response shows an in-depth and complete understanding of the topic.				
2. Relationships among science facts and concepts are clearly, completely, and accurately explained and fully supported with relevant data, examples, or citations.				
3. Ideas are expressed clearly and succinctly in a logical manner.				
4. All aspects of the question are adequately addressed.				
5. Spelling and language conventions are correctly applied.				
6. Language used in the response is appropriate for the needs of the audience.				

Comments	Goals	Actions

Figure 7.1   Brief Constructed Response to Open-Ended Science Questions (Performance List Rubric)

Brief Constructed Response to Open-Ended Science Questions (Holistic Rubric)

Name _____ Date _____ Course/Class _____

Task/Assignment _____

Expert 4	The response shows an in-depth understanding of the topic. Relationships among science facts and concepts are clearly, completely, and accurately explained and fully supported with relevant data, examples, or citations. Ideas are expressed clearly and succinctly in a logical manner. All aspects of the question are addressed. Spelling and language conventions are correctly applied. Language used in the response is appropriate for the needs of the audience.
Proficient 3	The response shows a solid understanding of the topic. Relationships among science facts and concepts are explained and generally supported with relevant data, examples, or citations. Ideas are expressed, for the most part, clearly and succinctly. The various aspects of the question are generally addressed. Spelling and language conventions are generally correct. Language used in the response is largely appropriate for the needs of the audience. Minor errors do not interfere with meaning.
Emergent 2	The response shows a partial understanding of the topic. There is an attempt to explain the relationships among science facts and concepts, but some serious omissions or misconceptions are evident. Insufficient support is provided. Ideas are not always expressed in a clear and logical manner, making the response difficult to follow. The question is only partially addressed. Flaws in spelling and language conventions interfere. Language used in the response is mostly inappropriate for the needs of the audience.
Novice 1	The response shows a very limited understanding of (or serious misconceptions about) the topic. Relationships among science facts and concepts are not explained. Little or no support is provided. Ideas are not presented in a clear and logical manner. The question is not completely or satisfactorily addressed. Major flaws in spelling and language conventions make the response difficult to follow. Language used in the response is inappropriate for the needs of the audience.

Comments	Goals	Actions

Figure 7.2 Brief Constructed Response to Open-Ended Science Questions (Holistic Rubric)

Classifying Objects or Processes (Performance List Rubric)

Name _____   Date _____   Course/Class _____

Task/Assignment _____

Performance Criteria	Assessment			
	Points	Self	Teacher	Other(s)
1. The classification system lists characteristics that can be observed directly and do not have to be inferred.				
2. The chosen characteristics are significant and clearly discriminate among the objects or processes being classified.				
3. The classification system is clear and logical.				
4. The characteristics of the chosen object or process begin with the most general (inclusive) and proceed to the most specific (discrete).				
5. The classification system accommodates new objects or processes related to the original set.				
6. The language chosen to define the attributes is scientifically accurate, descriptive, and useful.				

Comments	Goals	Actions

Figure 7.3   Classifying Objects or Processes (Performance List Rubric)

Classifying Objects or Processes (Holistic Rubric)

Name _____ Date _____ Course/Class _____

Task/Assignment _____

Expert 4	The classification system is thorough, detailed, flows naturally from general to more specific characteristics, and branches logically. The characteristics chosen for the system are clearly stated, do not overlap, and clearly discriminate among the objects or processes. Several people, using the same objects or processes, can use the system and get the same results. The system is flexible to the extent that all new objects or processes, which are related to the original set, can be classified. Language used in the classification system is scientifically accurate, descriptive, purposeful, and useful.
Proficient 3	The classification system is complete, and for the most part flows naturally from general to more specific characteristics, and branches logically. The characteristics chosen for the system are relevant, have little overlap, and discriminate among the objects or processes. Several people, using the same objects or processes, can use the system and get the same results most of the time. The system is flexible to the extent that most new objects or processes, which are related to the original set, can be classified. Language used in the classification system is scientifically accurate, descriptive, purposeful, and useful.
Emergent 2	The classification system contains significant omissions, flows somewhat unnaturally from general to more specific characteristics, and branches illogically. Some of the characteristics chosen for the system are irrelevant, have some overlap, and do not clearly discriminate among the objects or processes. Several people, using the same objects or processes, will experience some difficulty in using the system and getting the same results. The system is somewhat flexible, but it is difficult to classify new objects or processes, that are related to the original set. Language used in the classification system has some scientific inaccuracies, is not always descriptive, and lacks a clear purpose.
Novice 1	The classification system contains major significant omissions, does not flow from general to more specific characteristics, and branches illogically. Most of the characteristics chosen for the system are irrelevant, overlap, and do not discriminate among the objects or processes. Several people, using the same objects or processes, will get quite different results. The system lacks flexibility, making it impossible to classify new objects or processes, that are related to the original set. Language used in the classification system has many scientific inaccuracies, is not descriptive, and lacks a clear purpose.

Comments	Goals	Actions

Figure 7.4   Classifying Objects or Processes (Holistic Rubric)

Classifying Objects or Processes (Analytic Rubric)

Name _____ Date _____ Course/Class _____

Task/Assignment _____

	Characteristics	Order	Reliability	Language
Expert **(4)**	The characteristics chosen include those that are important and are clearly evident when using standard methods of examination. There is no confusion between or overlapping of characteristics chosen for inclusion in the system. The characteristics clearly discriminate among the objects or processes to be classified. The classification system lists characteristics that can be observed directly and do not have to be inferred. ☐	The characteristics of the object or process chosen begin with the most general, most inclusive (holistic) and logically proceed to the most specific. The classification system flows naturally, branches logically, and is easy to use. ☐	Several people can use the classification system with the same set of objects or processes and get the same results. The classification system is flexible to the extent that new objects or processes can be classified, which have not been previously classified but are related to the original set. ☐	The language chosen to define the attributes is scientifically accurate, descriptive, purposeful, and useful. ☐
Proficient **(3)**	The characteristics chosen include those that are important and are evident when using standard methods of examination. There is some confusion between or overlapping of characteristics chosen for inclusion in the system. The characteristics largely discriminate among the objects or processes to be classified. The classification system lists characteristics that can be observed directly and do not have to be inferred. ☐	The characteristics of the object or process chosen begin with the most general, most inclusive (holistic) and logically proceed to the most specific. For the most part, the classification system flows naturally, branches logically, and is easy to use. ☐	Several people can use the classification system with the same set of objects or processes and may not always get the same results. The classification system is flexible, but there is some difficulty in classifying new objects or processes, which have not been previously classified but are related to the original set. ☐	The language chosen to define the attributes is substantially accurate, descriptive, purposeful, and useful. ☐

Figure 7.5   Classifying Objects or Processes (Analytic Rubric)

(Continued)

Classifying Objects or Processes (Analytic Rubric) (continued)

	Characteristics	Order	Reliability	Language
Emergent (2)	The characteristics chosen include those that are somewhat important and are usually evident when using standard methods of examination. There is much confusion between or overlapping of characteristics chosen for inclusion in the system. The characteristics do not always discriminate among the objects or processes to be classified. The classification system lists characteristics that cannot always be observed directly and may have to be inferred. ❑	The characteristics of the object or process chosen do not always begin with the most general, most inclusive (holistic) and logically proceed to the most specific. For the most part, the classification system flows awkwardly, branches illogically, and is difficult to use. ❑	Several people can use the classification system with the same set of objects or processes and get quite different results. The classification system is inflexible; there is much difficulty in classifying new objects or processes, which have not been previously classified but are related to the original set. ❑	The language chosen to define the attributes is partially accurate, descriptive, purposeful, and useful. ❑
Novice (1)	The characteristics chosen include those that are trivial and are not evident when using standard methods of examination. There is so much confusion between or overlapping of characteristics chosen for inclusion in the system that the system becomes useless. The characteristics do not discriminate among the objects or processes to be classified. The classification system lists characteristics that rarely can be observed directly. ❑	The characteristics of the object or process chosen rarely begin with the most general, most inclusive (holistic) and logically proceed to the most specific. For the most part, there is no flow or order to the classification, making it impossible to use. ❑	Several people can use the classification system with the same set of objects or processes and get quite different results. The classification system is so inflexible that new objects or processes, which have not been previously classified but are related to the original set, cannot be reliably classified. ❑	The language chosen to define the attributes contains many inaccuracies, and lacks description, purpose, and use. ❑

Class Participation (Performance List Rubric)

Name _____   Date _____   Course/Class _____

Task/Assignment _____

Performance Criteria	Assessment			
	Points	Self	Teacher	Other(s)
1. Is alert, attentive, focused, and on-task.				
2. Follows established rules and procedures for class behavior.				
3. Contributes relevant and thoughtful information to class and cooperative group discussions.				
4. Respects the opinions of others and listens to classmates.				
5. Comes to class prepared – brings necessary materials, completes work, and turns in assignments on time.				
6. Is cooperative, willingly provides ideas for the group to consider, and follows established timelines.				

Comments   Goals   Actions

Figure 7.6   Class Participation (Performance List Rubric)

Class Participation (Analytic Rubric)

Name _____

Task/Assignment _____

Date _____   Course/Class _____

	Attention	Participation	Preparation	Group Work
Expert (4)	Always exhibits these behaviors: • Alertness • Becomes quiet and focused when asked • Faces in appropriate direction • Listens carefully to instructor and directions • Follows rules and procedures set by instructor ❏	Always: • Contributes relevant information • Contributes in a thoughtful manner • Contributes at appropriate times • Listens to classmates • Uses classroom and laboratory material efficiently and effectively • Follows directions ❏	Always: • Brings book, pencil/pen, and notebook • Completes assignments • Hands in work on time ❏	Always: • Is cooperative • Is willing to work with group • Is willing to provide ideas • Listens to group members • Is aware of time limits and deadlines ❏
Proficient 3	Often exhibits these behaviors: • Alertness • Becomes quiet and focused when asked • Faces in appropriate direction • Listens to instructor and directions • Follows rules and procedures set by instructor ❏	Often: • Contributes relevant information • Contributes in a thoughtful manner • Contributes at appropriate times • Listens to classmates • Uses classroom and laboratory material efficiently and effectively • Follows directions ❏	Often: • Brings book, pencil/pen, and notebook • Completes assignments • Hands in work on time ❏	Often: • Is cooperative • Is willing to work with group • Is willing to provide ideas • Listens to group members • Is aware of time limits and deadlines ❏

Figure 7.7   Class Participation (Analytic Rubric)

Class Participation (continued)

	Attention	Participation	Preparation	Group Work
Emergent **(2)**	Occasionally exhibits these behaviors: • Alertness • Becomes quiet and focused when asked • Faces in appropriate direction • Listens to instructor and directions • Follows rules and procedures set by instructor ❏	Occasionally: • Contributes relevant information • Contributes in a thoughtful manner • Contributes at appropriate times • Listens to classmates • Uses classroom and laboratory material efficiently and effectively • Follows directions ❏	Occasionally: • Brings book, pencil/pen, and notebook • Completes assignments • Hands in work on time ❏	Occasionally: • Is cooperative • Is willing to work • Is willing to provide ideas • Listens to group members • Is aware of time limits and deadlines ❏
Novice **(1)**	Rarely exhibits these behaviors: • Alertness • Becomes quiet and focused when asked • Faces in appropriate direction • Listens to instructor and directions • Follows rules and procedures set by instructor ❏	Rarely: • Contributes relevant information • Contributes in a thoughtful manner • Contributes at appropriate times • Listens to classmates • Uses classroom and laboratory material efficiently and effectively • Follows directions ❏	Rarely: • Brings book, pencil/pen, and notebook • Completes assignments • Hands in work on time ❏	Rarely: • Is cooperative • Is willing to work • Is willing to provide ideas • Listens to group members • Is aware of time limits and deadlines ❏

Concept Mapping (Performance List Rubric)

Name _____ Date _____ Course/Class _____

Task/Assignment _____

<table>
<tr><td rowspan="2">Performance Criteria</td><td colspan="4" align="center">Assessment</td></tr>
<tr><td>Points</td><td>Self</td><td>Teacher</td><td>Other(s)</td></tr>
<tr><td>1. The map presents a definite hierarchical arrangement, with the most general, inclusive concept at the top and the most specific concepts at the bottom.</td><td></td><td></td><td></td><td></td></tr>
<tr><td>2. Relevant and scientifically accurate linking words clearly connect all the concepts.</td><td></td><td></td><td></td><td></td></tr>
<tr><td>3. Concepts that are equivalent in specificity are placed at the same level (same hierarchy) within the concept map.</td><td></td><td></td><td></td><td></td></tr>
<tr><td>4. Crosslinks between concepts are made to further explain the relationships.</td><td></td><td></td><td></td><td></td></tr>
<tr><td>5. The concept map has an appropriate title that specifies the main idea.</td><td></td><td></td><td></td><td></td></tr>
<tr><td>6. The concept map is logical and easy to understand.</td><td></td><td></td><td></td><td></td></tr>
</table>

Comments	Goals	Actions

Figure 7.8   Concept Mapping (Performance List Rubric)

Concept Mapping (Holistic Rubric)

Name _____   Date _____   Course/Class _____

Task/Assignment _____

Expert 4	The map presents a definite hierarchical arrangement, with the most general, inclusive concept at the top and the most specific concepts at the bottom. Relevant and scientifically accurate linking words clearly connect all the concepts. Concepts that are equivalent in specificity are placed at the same level (same hierarchy) within the concept map. Crosslinks between concepts are made to further explain the relationships. The concept map has an appropriate title that specifies the main idea. The concept map is logical and easy to understand.
Proficient 3	For the most part the map presents a definite hierarchical arrangement. Relevant and scientifically accurate linking words connect all the concepts. Concepts that are equivalent in specificity are, with minor errors, placed at the same level (same hierarchy) within the concept map. Crosslinks between concepts are stated, but do not contribute significantly to further explanations between the relationships. The concept map has an appropriate title that specifies the main idea. The concept map is mostly logical and easy to understand.
Emergent 2	The map presents a hierarchical arrangement, but major misconceptions exist about which concepts are more inclusive and which are more specific. Some linking words are either missing, scientifically inaccurate, or irrelevant. Concepts that are equivalent in specificity are scattered throughout the map. The concept map has a title that does little to clarify the main idea. The concept map is awkard to understand.
Novice 1	There is no hierarchical arrangement to the map. Inclusive and specific concepts are mixed throughout. Linking words are missing, contain major scientific inaccuracies, and are irrelevant. There is no pattern to the placement of concepts that are equivalent in specificity. The concept map has no title and is difficult to understand.

Comments	Goals	Actions

Figure 7.9   Concept Mapping (Holistic Rubric)

Cooperative Learning (Performance List Rubric)

Name _____ Date _____ Course/Class _____

Task/Assignment _____

Performance Criteria	Assessment			
	Points	Self	Teacher	Other(s)
1. My group completed all that was expected.				
2. I accomplished my assigned task within the group.				
3. My actions and behavior made significant contributions to the total group effort.				
4. I listened to and respected others and their ideas.				
5. I got others involved by asking questions, challenging, and/or requesting information.				
6. Working in the group helped me learn more than if I had worked alone.				

Comments	Goals	Actions

Figure 7.10   Cooperative Learning (Performance List Rubric)

Data Interpretation (Performance List Rubric)

Name _____   Date _____   Course/Class _____

Task/Assignment _____

Performance Criteria	Assessment			
	Points	Self	Teacher	Other(s)
1. The interpretation includes all relevant data elements.				
2. Appropriate statistical procedures were selected and applied.				
3. The formula(s) for the statistical interpretation is/are correctly applied to solve for the unknown quantity.				
4. All mathematical calculations are accurate.				
5. Appropriate graphic display techniques were used (in addition to the mathematical formulas) to further analyze the data.				
6. The interpretation revealed any significant patterns in the data.				
7. Appropriate inferences and/or conclusions were made based on the data interpretation.				

Comments	Goals	Actions

Figure 7.11   Data Interpretation (Performance List Rubric)

Data Interpretation (Holistic Rubric)

Name _____ Date _____ Course/Class _____

Task/Assignment _____

Expert 4	The interpretation includes all data elements with an adequate sample size and repeated trials being used. Appropriate statistical procedures were selected and applied. The formula(s) for the statistical interpretation is/are rearranged correctly to solve for the unknown quantity. Existing patterns can be discerned within the data, based upon the statistical interpretation. All mathematical calculations are accurate. Appropriate graphic techniques were used to further display and analyze the data.
Proficient 3	The interpretation includes most data elements with an adequate sample size and repeated trials being used. Statistical procedures were selected and applied. The formula(s) for the statistical interpretation is/are rearranged with minor errors that do not affect the outcome. Existing patterns can be discerned within the data, based upon the statistical interpretation. Minor errors may exist in the mathematical calculations. Graphic techniques, on a limited basis, were used in addition to the mathematical formulas.
Emergent 2	The interpretation is missing significant data elements. Sample size is not large enough to extrapolate to the population. Repeated trials were used on a limited basis to generate data for interpretation. Statistical procedures were haphazardly selected and applied. It is difficult to discern existing patterns within the data, based upon the statistical interpretation. Major errors exist in the mathematical calculations. Some graphic techniques were attempted, in addition to the mathematical formulas.
Novice 1	The interpretation is missing most data elements. Sample size is not addressed. Attempts to use repeated trials are evident, but with little understanding of their value or application to data interpretation. Statistical procedures were not identified nor applied. It is impossible to discern existing patterns within the data, based upon the statistical interpretation. Major errors exist in the mathematical calculations. No graphic techniques were attempted, in addition to the mathematical formulas to further display and analyze the data.

Comments	Goals	Actions

Figure 7.12 Data Interpretation (Holistic Rubric)

Data Interpretation (Analytic Rubric)

Name _____ Date _____ Course/Class _____

Task/Assignment _____

	Sample Size/Repeated Trials	Algebraic Analysis/Data Interpretation	Graphic Representation of Data
Expert **(4)**	The interpretation includes all data elements, not just selected data sets. An adequate sample size and repeated trials were used to generate data for interpretation. ☐	Appropriate statistical procedures were selected and applied. The formula(s) for the statistical interpretation is/are rearranged correctly to solve for the unknown quantity. Existing patterns can be discerned within the data based upon the statistical interpretation. All mathematical calculations are accurate. Extensive data interpretation was conducted to allow for the drawing of inferences and conclusions. ☐	Appropriate graphic techniques were used to further display and analyze the data. ☐
Proficient **(3)**	The interpretation includes most data elements. An adequate sample size and repeated trials were used to generate data for interpretation. ☐	Statistical procedures were selected and applied. The formula(s) for the statistical interpretation is/are rearranged, with minor errors that do not affect the outcome, to solve for the unknown quantity. Existing patterns can be discerned within the data based upon the statistical interpretation. Minor errors exist in the mathematical calculations. Sufficient data interpretation was conducted to allow for the drawing of inferences and conclusions. ☐	Graphic techniques, on a limited basis, were used to further display and analyze the data. ☐

Figure 7.13 Data Interpretation (Analytic Rubric)

(Continued)

113

Data Interpretation (continued)

	Sample Size/Repeated Trials	Algebraic Analysis/Data Interpretation	Graphic Representation of Data
Emergent (2)	The interpretation is missing significant data elements. Sample size is not large enough to extrapolate to the population. Repeated trials were used on a limited basis to generate data for interpretation. ❏	Statistical procedures were haphazardly selected and applied. The formula(s) for the statistical interpretation is/are rearranged, with major errors that do not affect the outcome, to solve for the unknown quantity. It is difficult to discern existing patterns within the data based upon the statistical interpretation. Major errors exist in the mathematical calculations. Some data interpretation was conducted to allow for the drawing of inferences and conclusions. ❏	Some graphic techniques were attempted to further display and analyze the data. ❏
Novice (1)	The interpretation is missing most data elements. Sample size is not addressed. Attempts to use repeated trials are evident, but with little understanding of their value or application to data interpretation. ❏	Statistical procedures were not identified nor applied. It is impossible to discern existing patterns within the data based upon the statistical interpretation. Major errors exist in the mathematical calculations. On a very limited basis, data interpretation was conducted to allow for the drawing of inferences and conclusions. ❏	No graphic techniques were attempted to further display and analyze the data. ❏

114

Data Organization Table (Performance List Rubric)

Name _____ Date _____ Course/Class _____

Task/Assignment _____

Performance Criteria	Assessment			
	Points	**Self**	**Teacher**	**Other(s)**
1. The design of the table is appropriate for the types and quantities of data being collected.				
2. All relevant data are accurately and completely recorded in the table.				
3. The data organization table has a title that reflects the relationship between the independent and dependent variables.				
4. A key or legend for the table is provided (if needed).				
5. The information in the columns and rows is appropriately organized and correctly labeled.				
6. The cells within the table are appropriately scaled and consistent in size.				
7. The set of data was recorded within the appropriate cells.				
8. All measurements are labeled with the correct magnitude (numerical value) using metric units.				
9. The data organization table is complete, attractive, and presentable.				
Optional				
10. The data organization table has been constructed with computer generated or other graphic tools.				
11. Data from multiple trials are clearly shown.				

Comments	Goals	Actions

Figure 7.14 Data Organization Table (Performance List Rubric)

Data Organization Table (Holistic Rubric)

Name _____ Date _____ Course/Class _____

Task/Assignment _____

Expert 4	The table is expertly designed for the types and quantities of data being collected. All relevant data are accurately and completely recorded in the table, with a key or legend for the table, and a title that clearly reflects the relationship between the independent and dependent variables is provided. The data table columns and rows are appropriately organized, labeled, and scaled with the data fitting completely and clearly within the cells. All measurements are labeled with the correct magnitude (numerical value), metric unit, and have an appropriate number of significant figures. The data table is neat, attractive, and presentable.
Proficient 3	The table is appropriate for most of the types and quantities of data being collected. All relevant data are recorded, with only minor errors, in the table. A key or legend for the table and a title that reflects the relationship between the independent and dependent variables are provided. The data table columns and rows are appropriately organized, labeled, and scaled with the data fitting completely and clearly within the cells. Most measurements are labeled with the correct magnitude (numerical value) and metric unit. The data table is neat, attractive, and presentable.
Emergent 2	The table is appropriate for some of the types and quantities of data being collected. There is an attempt to identify the independent and dependent variables, but they are not clearly defined within the table or in the title. The data organization table is inappropriately sized, labeled, and organized. Some measurements are labeled with the correct magnitude (numerical value) and metric unit. The data table lacks neatness.
Novice 1	Major flaws exist in the design of the data organization table. The identification of variables is poorly done and lacks clarity. The size and structure of the table is inappropriate for the types and quantities of data being collected. Data columns and rows are not accurately labeled nor completed. Data units are missing. The data organization table is neither neat nor presentable.

Comments	Goals	Actions

Figure 7.15   Data Organization Table (Holistic Rubric)

Data Organization Table (Analytic Rubric)

Name _____

Task/Assignment _____

Date _____ Course/Class _____

	Organization of Table	Entering the Data Into the Table	Clarity and Neatness of Table
Expert (4)	The data organization table is expertly designed either by computer or graphic tools and is appropriate for the types and quantities of data being collected. A key or legend for the table is provided if needed to interpret the data. Both the independent and dependent variables are clearly defined within the table and in the title. The cells within the table are appropriately scaled, consistent in size, with the data fitting completely and clearly within the cells. ☐	The information in the data table columns and rows is appropriately organized and labeled. The set of data was recorded within the appropriate cells. All measurements are labeled with the correct magnitude (numerical value) and metric unit. The data have an appropriate number of significant figures. The data are amazingly accurate based upon the measuring equipment or instrument being used. Data from multiple trials at each level of the independent variable are clearly shown. ☐	The data table is complete, neat, attractive, and graphically appealing. ☐
Proficient (3)	The design of the data organization table is appropriate for the types and quantities of data being collected. Both the independent and dependent variables are clearly defined within the table and in the title. The data organization table is appropriately sized and clearly displays all the collected data. ☐	The information in the data table columns is appropriately organized and labeled. The set of data was recorded within the appropriate cells. All measurements are labeled with the correct magnitude (numerical value) and metric unit. The data have an appropriate number of significant figures. Accuracy of the data is appropriate to the measuring equipment or instrument being used. Data from multiple trials at each level of the independent variable are clearly shown. ☐	The data table is neat and presentable. ☐

Figure 7.16 Data Organization Table (Analytic Rubric)

(Continued)

117

Data Organization Table (continued)

	Organization of Table	Entering the Data Into the Table	Clarity and Neatness of Table
Emergent **(2)**	The design of the data organization table is somewhat appropriate for the types and quantities of data being collected. There is some confusion about the independent and dependent variables. The data organization table is inappropriately sized and is missing some of the collected data. ❐	The information in the data table columns lacks organization and some labeling. Some of the data were recorded within the appropriate cells. Some measurements are labeled with the correct magnitude (numerical value) and metric unit. Some data have an appropriate number of significant figures. The accuracy of the data could be improved with the measuring equipment or instrument being used. Multiple trials at each level of the independent variable are lacking. ❐	The data table is neat and presentable. ❐
Novice **(1)**	Major flaws exist in the design of the data organization table. The identification of variables is unclear. The size and structure of the table are not appropriate for the types and quantities of data being collected. ❐	Data columns and rows are not accurately labeled nor comprehensively completed. Data units are missing. ❐	Data organization table is neither neat nor presentable. ❐

Formulating Questions for Further Study (Performance List Rubric)

Name _____ Date _____ Course/Class _____

Task/Assignment _____

Performance Criteria	Assessment			
	Points	Self	Teacher	Other(s)
1. The question is important enough to warrant additional time and study.				
2. The question is developed from and is in response to background readings, data interpretation, observations, or information presented in the classroom.				
3. The question is an attempt to construct and connect new knowledge to prior knowledge.				
4. The question contains enough details to clearly define the reasoning for asking it.				
5. Scientific vocabulary appropriate for framing the question is used.				
6. The question is clearly stated in unambiguous language.				
7. The question is written or stated in a complete interrogative sentence.				
8. Correct language usage, punctuation, spelling and capitalization are used to frame the question.				
Optional 9. The question being formulated is testable (i. e., the question could be answered through a scientific investigation).				

Comments	Goals	Actions

Figure 7.17   Formulating Questions for Further Study (Performance List Rubric)

Graphic Organizer (Performance List Rubric)

Name _____ Date _____ Course/Class _____

Task/Assignment _____

Performance Criteria	Assessment			
	Points	**Self**	**Teacher**	**Other(s)**
1. An appropriate type of graphic organizer was used to represent the science concepts or processes.				
2. Conceptual information that is displayed within the organizer is scientifically accurate.				
3. Appropriately sized geometric shapes are used throughout to clearly represent science concepts or processes.				
4. Relationships among the geometric shapes are clearly shown with connecting lines.				
5. The eye of the observer is immediately drawn to the topic and main supporting concepts.				
6. There is a natural flow and order to the graphic organizer that allows for easy interpretation by others.				
7. A variety of graphic features such as different textures, shapes, and colors are used to highlight information and enhance the organizer's effectiveness.				
8. Relationships shown among the concepts are accurate and relevant to the topic.				
9. The organizer is neatly drawn, legible, and attractive.				

Comments	Goals	Actions

Figure 7.18   Graphic Organizer (Performance List Rubric)

Graphing Scientific Data (Performance List Rubric)

Name _____ Date _____ Course/Class _____

Task/Assignment _____

Performance Criteria	Assessment			
	Points	Self	Teacher	Other(s)
1. An appropriate type of graph was expertly used (line graph, bar graph, pictograph, histogram, stem-and-leaf, circle graph, line plot, etc.) to display the data set(s).				
2. The title of the graph clearly identifies the data displayed on the graph.				
3. The range of data was used to choose an appropriate sequence of numbers for both the x-and the y-axes (2's, 3's, 5's, 10's, 100's, etc.).				
4. Physical intervals on the graph are scaled appropriately and spaced evenly.				
5. All the parts of the graph are clearly labeled (units of measurement, x and y-axes, columns, rows, etc.).				
6. The independent (manipulated) variable is labeled on the x-axis and the dependent (responding) variable is labeled on the y-axis.				

Figure 7.19   Graphing Scientific Data (Performance List Rubric)

(Continued)

Graphing Scientific Data (continued)

Performance Criteria	Assessment			
	Points	**Self**	**Teacher**	**Other(s)**
7. A very precise technique is used to plot the data points.				
8. The graph can be used for predictive purposes.				
9. If needed, a key is provided.				
10. Colors, textures, labels, graphics, or other features are used to enhance the graph.				
11. The graph is clear and complete.				

Comments	Goals	Actions

Graphing Scientific Data (Holistic Rubric)

Name _____   Date _____   Course/Class _____

Task/Assignment   _____

Expert 4	An appropriate type of graph was expertly used for the data set(s). The title of the graph clearly relates to the data displayed and reflects both the independent and dependent variables. Physical intervals on the graph are scaled appropriately and spaced evenly. All the parts of the graph are clearly and accurately labeled. The set of data is plotted on the graph completely and accurately and the slope of the relationship is indicated. Colors, textures, labels, or other features are used to enhance the graph.
Proficient 3	An appropriate type of graph was used for the data set(s). The title of the graph relates to the data displayed and reflects both the independent and dependent variables. Physical intervals on the graph are scaled appropriately and spaced evenly. Most parts of the graph are clearly and accurately labeled. The set of data is plotted with only minor errors. Colors, textures, labels, or other features are used to enhance the graph.
Emergent 2	An appropriate type of graph was used for the data set(s). The title of the graph relates somewhat to the data displayed, but does not reflect both the independent and dependent variables. Physical intervals on the graph are scaled appropriately and spaced evenly. Some confusion exists as to labeling the parts of the graph. The set of data is plotted with some errors. There is minimal use of colors, textures, labels, or other features to enhance the graph.
Novice 1	An inappropriate type of graph was used for the data set(s). The title of the graph vaguely relates to the data displayed and does not reflect both the independent and dependent variables. Major problems exist with labeling the axes with an appropriate sequence of numbers based upon the range of the data. Physical intervals on the graph are not scaled appropriately nor spaced evenly. Much confusion exists as to labeling the parts of the graph. The set of data is plotted, with many errors. There is little, if any, use of colors, textures, labels, or other features to enhance the graph.

Comments	Goals	Actions

Figure 7.20   Graphing Scientific Data (Holistic Rubric)

Name _____ Date _____ Course/Class _____

Task/Assignment _____

Graphing Scientific Data (Analytic Rubric)

	Organization of Graph(s)	Display of Variables	Accuracy of Plotting of Data	Clarity and Neatness of Graph
Expert (4)	An appropriate type of graph was expertly used (line graph, bar graph, pictograph, histogram, stem-and-leaf, circle graph, line plot, etc.). The title of the graph very clearly relates to the data displayed on the graph. The range of data was used to choose an appropriate consistent sequence of numbers to number the x axis and the y axis (2's, 3's, 5's, 10's, 100's, etc.) Physical intervals on the graph are scaled appropriately and spaced evenly. All the parts of the graph are labeled (units of measurement, x and y axes, columns, rows, etc.). A key or legend was added to the graph to further explain the data. ❑	The independent (manipulated) variable is placed on the x-axis and the dependent (responding) variable is placed on the y-axis. ❑	A very precise technique is used to plot the data points. All data points are accurately plotted and the slope of the relationship is indicated. ❑	The graph clearly shows the relationship between the independent and dependent variables. The graph can be used for predictive purposes. Colors, textures, labels, or other features are used to enhance the graph. The graph is clear and complete. ❑
Proficient (3)	An appropriate type of graph was used (line graph, bar graph, pictograph, histogram, stem-and-leaf, circle graph, line plot, etc.). The title of the graph clearly relates to the information displayed on the graph. The data were used to choose an appropriate consistent interval to number the x axis and y axis (2's, 3's, 5's, 10's, 100's, etc.). Numbers on the graph are spaced evenly. All the parts of the graph are labeled (units of measurement, x and y axes, columns, rows, etc.). ❑	The independent (manipulated) variable is placed on the x-axis and the dependent (responding) variable is placed on the y-axis. ❑	The set of data is plotted on the graph completely and accurately. ❑	The graph clearly shows the relationship between the independent and dependent variables. The graph can be used for predictive purposes. The graph is clear and complete. ❑

Figure 7.21 Graphing Scientific Data (Analytic Rubric)

Graphing Scientific Data (continued)

	Organization of Graph(s)	Display of Variables	Accuracy of Plotting of Data	Clarity and Neatness of Graph
Emergent (2)	The response is generally like a 3, but the components are not as well defined and need additional work. Minor revisions would suffice to score a 3. ❑	The response is generally like a 3, but the components are not as well defined and need additional work. Minor revisions would suffice to score a 3. ❑	The response is generally like a 3, but the components are not as well defined and need additional work. Minor revisions would suffice to score a 3. ❑	The response is generally like a 3, but the components are not as well defined and need additional work. Minor revisions would suffice to score a 3. ❑
Novice (1)	An inappropriate type of graph was used (line graph, bar graph, pictograph, histogram, stem-and-leaf, circle graph, line plot, etc.). The title of the graph vaguely relates to the information displayed on the graph. The interval used to plot the data is not appropriate. Few parts of the graph are labeled (units of measurement, x and y axes, columns, rows, etc.). ❑	The variables are placed on the wrong axes. ❑	The set of data is plotted on the graph in a haphazard and imprecise way. ❑	The graph is unclear and incomplete. ❑

Hypothesis Testing (Performance List Rubric)

Name _____ Date _____ Course/Class _____

Task/Assignment _____

Performance Criteria	Assessment			
	Points	Self	Teacher	Other(s)
1. The hypothesis clearly guides the development of an experiment of the question.				
2. The hypothesis is based upon extensive background research, observations, and/or what is already known about the question.				
3. The hypothesis links the effect to the independent variable.				
4. The directionality of the effect is stated.				
5. The hypothesis states what will be the expected effect/change.				
6. Both the dependent and independent variables are identified within the hypothesis statement.				
7. The hypothesis is written in the "If-and- then" format. Example: **If** baking soda is added to the water **and** the *Elodea* is placed in front of a strong light source, **then** the rate of photosynthesis will increase.				
8. The hypothesis is written as a clear declarative sentence and the language used is purposeful and appropriate for the audience.				

Comments	Goals	Actions

Figure 7.22   Hypothesis Testing (Performance List Rubric)

Hypothesis Testing (Holistic Rubric)

Name _____   Date _____   Course/Class _____

Task/Assignment _____

Expert 4	The hypothesis clearly guides the development of an experiment for the research question and is based upon extensive background research, observations, and/or what is already known about the question. The hypothesis directly links the effect to the independent variable, with the directionality of the effect being stated. The hypothesis states what will be the expected effect/change. Both the dependent and independent variables are identified within the hypothesis statement. The hypothesis is written in the "If-and-then" format. The hypothesis is written as a clear declarative sentence and the language used is purposeful and appropriate for the audience.
Proficient 3	The hypothesis guides the development of an experiment of the research question and is based upon background research, observations, and/or what is already known about the question. The hypothesis links the effect to the independent variable, with the directionality of the effect being stated. The hypothesis states what will be the expected effect/change. Both the dependent and independent variables are identified within the hypothesis statement. The hypothesis is written in the "If-and-then" format. The hypothesis is written as a clear declarative sentence.
Emergent 2	The hypothesis provides some guidance in the development of an experiment of the research question and is based upon limited background research, observations, and/or what is already known about the question. The hypothesis attempts to link the effect to the independent variable, but with limited success. The directionality of the effect is unclear. The hypothesis vaguely states what will be the expected effect/change. There is an attempt to link the dependent and independent variables within the hypothesis statement. The hypothesis is not written in the "If-and-then" format. The hypothesis is written as a clear declarative sentence.
Novice 1	The hypothesis provides little to no guidance in the development of an experiment of the research question. The hypothesis is based upon little to no background research, observations, and/or what is already known about the question. The hypothesis attempts to link the effect to the independent variable, but with little success. The directionality of the effect is not stated. The hypothesis does not specify what will be the expected effect/change. There is some attempt, but little success, to link the dependent and independent variables within the hypothesis statement. The hypothesis is not written in the "If-and-then" format. The hypothesis is not written as a complete declarative sentence.

Comments	Goals	Actions

Figure 7.23   Hypothesis Testing (Holistic Rubric)

Hypothesis Testing (Analytic Rubric)

Name _____ Date _____ Course/Class _____

Task/Assignment _____

	Guidance for Formation of Hypothesis	Linkage and Directionality	Form of Hypothesis	Language Usage
Expert (4)	The hypothesis clearly guides the development of an experiment of the research question. The hypothesis is based upon extensive background research, observations, and/or what is already known about the question. ❑	The hypothesis directly and clearly links the effect to the independent variable. The directionality of the effect is stated. The hypothesis states what will be the expected effect/change. Both the dependent and independent variables are identified within the hypothesis statement. ❑	The hypothesis is written in the "If-and-then" format. Example: **If** baking soda is added to the water **and** the <u>Elodea</u> is placed in front of a strong light source, **then** the rate of photosynthesis should increase. ❑	The hypothesis is written as a clear declarative sentence and the language used is purposeful and appropriate for the audience. ❑
Proficient (3)	The hypothesis guides the development of an experiment of the research question. The hypothesis is based upon background research, observations, and/or what is already known about the question. ❑	The hypothesis links the effect to the independent variable. The hypothesis states what will be the expected effect/change. Both the dependent and independent variables are identified within the hypothesis statement. ❑	The hypothesis is written in the "If-and-then" format. Example: **If** baking soda is added to the water **and** the <u>Elodea</u> is placed in front of a strong light source, **then** the rate of photosynthesis should increase. ❑	The hypothesis is written as a clear declarative sentence. ❑

Figure 7.24 Hypothesis Testing (Analytic Rubric)

Hypothesis Testing (continued)

	Guidance for Formation of Hypothesis	Linkage and Directionality	Form of Hypothesis	Language Usage
Emergent (2)	The hypothesis provides some guidance in the development of an experiment of the research question. The hypothesis is based upon limited background research, observations, and/or what is already known about the question. ❑	The hypothesis attempts to link the effect to the independent variable, but with limited success. The directionality of the effect is unclear. The hypothesis vaguely states what will be the expected effect/change. There is an attempt to link the dependent and independent variables within the hypothesis statement. ❑	The hypothesis is not written in the "If-and-then" format. ❑	The hypothesis is written as a clear declarative sentence. ❑
	❑	❑	❑	❑
Novice (1)	The hypothesis provides little to no guidance in the development of an experiment of the research question. The hypothesis is based upon little or no background research, observations, and/or what is already known about the question. ❑	The hypothesis attempts to link the effect to the independent variable, but with no success. The directionality of the effect is not stated. The hypothesis does not specify what will be the expected effect/change. There is some attempt, but little success, to link the dependent and independent variables within the hypothesis statement. ❑	The hypothesis is not written in the "If-and-then" format. ❑	The hypothesis is not written as a complete declarative sentence. ❑
	❑	❑	❑	❑

Journal in Science (Performance List Rubric)

Name _____ Date _____ Course/Class _____

Task/Assignment _____

Performance Criteria	Assessment			
	Points	Self	Teacher	Other(s)
1. The cover of the journal contains a descriptive title, the name of the owner, the subject or class, and the dates the journal covers.				
2. Each entry in the journal contains a descriptive title, the date(s) of the entry, and documentation of the science content and processes.				
3. Journal entries follow a set pattern that is consistent, well organized, and logical.				
4. Tables, charts, graphs, drawings, and other graphics are clearly titled and labeled and are regularly used.				
5. Entries in the journal are detailed enough that one could use them for study notes or open book tests.				
6. Rules of grammar and language usage are followed in developing and completing entries in the journal.				
Optional				
7. Self-reflection narratives are included with many entries.				
8. Within the self-reflection narratives are included: problems and concerns, questions still unanswered, issues to be resolved, and "what-if" questions.				

Comments	Goals	Actions

Figure 7.25   Journal in Science (Performance List Rubric)

Lab Report (Performance List Rubric)

Name _____ Date _____ Course/Class _____

Task/Assignment _____

Performance Criteria	Assessment			
	Points	Self	Teacher	Other(s)
Introduction to Report				
1. The title states clearly both the independent and dependent variables and the results of the experiment.				
2. The title of the report is written in a clear declarative statement.				
3. A concise abstract (not more than 250 words) of the lab is provided.				
Question/Problem				
4. The question/problem that the lab was designed to answer is clearly stated.				
5. Relevant literature and prior observations are cited.				
6. The hypothesis is stated in the "If-and-then" format. It predicts the influence of the independent variable on the dependent variable.				
Procedures for Experiment				
7. The procedures for controlling and measuring the dependent variable are well defined and clear.				
8. A detailed, logical, step-by-step set of procedures that were for conducting the lab is listed.				

Figure 7.26 Lab Report (Performance List Rubric)

(Continued)

Lab Report (continued)

Performance Criteria	Assessment			
	Points	Self	Teacher	Other(s)
9. Safety concerns are listed among the procedures.				
Data Organization and Display 10. Refer to "Data Organization Table" Assessment/Evaluation Tool.				
11. Refer to "Graphing Scientific Data" Assessment/Evaluation Tool.				
Data Analysis 12. Refer to "Data Interpretation" Assessment/Evaluation Tool.				
Conclusions 13. A response to both the question and hypothesis is clearly and completely provided and is consistent with the data.				
14. Interpretations, as well as limitations, of the data are included. 15. Unresolved questions and problems are listed.				
16. Questions for further study are developed.				
Language Usage 17. Language is used correctly and purposefully. 18. All words are spelled correctly.				
19. The report is neat, legible, and presentable.				

Comments	Goals	Actions

Lab Report (Analytic Rubric)

Name _____   Date _____   Course/Class _____

Task/Assignment _____

	Title/ Introduction	Background Research	Question/Problem/ Hypothesis	Procedures	Data & Results	Conclusions	Language Usage
Expert (4)	The title states clearly both the independent and dependent variables and the results of the experiment. The title of the report is written in a clear declarative statement. The lead-in information is concise and develops a clear understanding of the report to follow. A concise abstract of the lab is provided and does not exceed 250 words. ☐	Relevant literature and prior observations are cited which provide much insight into the phenomena to be included in the report. ☐	The question or problem that the lab was designed to answer is well articulated. The hypothesis is eloquently stated in the "If-and-then" format. It predicts the influence of the independent variable on the dependent variable. ☐	The procedures for controlling and measuring the dependent variable are well defined and clear. A detailed, logical step-by-step set of procedures that was used for conducting the lab is listed. Safety concerns are listed among the procedures. ☐	Data tables and graphs are expertly and neatly completed and totally accurate. Patterns or trends in data are noted. Data analysis is thorough. ☐	A response to both the question and hypothesis is clearly and completely provided and is consistent with the data. Limitations and extrapolations of the data are cited. Questions for further study are developed. Unresolved questions and problems are listed. ☐	Language is used correctly and purposefully. All words are spelled correctly. The report is neat, legible, and presentable. ☐
Proficient 3	The title states both the independent and dependent variables and the results of the experiment. The title of the report is written in a clear declarative statement. The lead-in information is concise and develops a clear understanding of the report to follow. ☐	Relevant literature and prior observations are cited which provide insight into the phenomena to be included in the report. ☐	The question or problem that the lab was designed to answer is listed. The hypothesis is stated in the "If-and-then" format. It predicts the influence of the independent variable on the dependent variable. ☐	The procedures for controlling and measuring the dependent variable are defined and clear. A detailed, logical step-by-step set of procedures that was used for conducting the lab is listed. Safety concerns are missing from the procedures. ☐	Data tables and graphs are neatly completed and totally accurate. Patterns or trends in data are noted. Data analysis is thorough. ☐	A response to both the question and hypothesis is provided. Some limitations and extrapolations of the data are cited. ☐	Language is used correctly and purposefully. Some words are misspelled, but with little or no effect upon the final product. The report is neat, legible, and presentable. ☐

Figure 7.27   Lab Report (Analytic Rubric)

133

(Continued)

Lab Report (continued)

	Title/Introduction	Background Research	Question/Problem/Hypothesis	Procedures	Data & Results	Conclusions	Language Usage
Emergent 2	The title is stated in a rambling, non-concise fashion. There is an attempt within the title to state both the independent and dependent variables and the results of the experiment. The title of the report is written in a declarative statement. The lead-in information lacks conciseness and clarity. ☐	Literature and prior observations are cited but provide little insight into the phenomena to be included in the report. ☐	The question or problem that the lab was designed to answer is ill defined. The hypothesis is stated, but not in the "If-and-then" format. ☐	Some of the steps are understandable; most are confusing and lack detail. ☐	Data tables and graphs are completed but lack accuracy. Patterns or trends within the data are difficult to discern. Data analysis lacks thoroughness. ☐	Presents an illogical explanation for findings. ☐	For the most part, language is used correctly. However, many words are misspelled, impacting upon the final product. The report borders on being sloppy, illegible, and not presentable. ☐
Novice 1	The title is stated in a rambling, non-concise fashion. There is no attempt within the title to state the independent and dependent variables and the results of the experiment. The title of the report is written in a declarative statement. The lead-in information provides little or no information that leads into the report. ☐	Literature and prior observations are not cited. ☐	The question or problem that the lab was designed to answer is not defined. There is no hypothesis. ☐	Steps are not sequential; most steps are missing or are confusing. ☐	Data tables and/or graphs are missing information and are inaccurate. Consequently, patterns or trends within the data are not discernable. Little attempt is made at data analysis. ☐	Presents an illogical explanation for findings and does not address the question that guided the lab. ☐	Language is used incorrectly and without purpose. Many words are misspelled, impacting significantly upon the final product. The report is definitely sloppy, illegible, and not presentable. ☐

Language in Use in Science Writing (Performance List Rubric)

Name _____ Date _____ Course/Class _____

Task/Assignment _____

Performance Criteria	Assessment			
	Points	Self	Teacher	Other(s)
1. Throughout the entire text there is evidence of the writer using language choices, including scientific vocabulary, for effective expression of meaning.				
2. The text uniformly conveys an impression of correctness and scientific accuracy with few, if any, errors. Errors that occur may appear as a consequence of risk-taking in language use or an elaboration upon a science concept or process.				
3. The writer consistently uses varied sentence formation to create style and tone and to enhance meaning.				
4. The writer consistently uses language, including scientific vocabulary, to create style and tone and to enhance meaning.				
5. The writer consistently demonstrates correct language usage, punctuation, spelling, and capitalization.				

Comments	Goals	Actions

Figure 7.28   Language in Use in Science Writing (Performance List Rubric)

Language in Use in Science Writing (Holistic Rubric)

Name _____ Date _____ Course/Class _____

Task/Assignment _____

Expert 4	Throughout the entire text there is evidence of the writer using language choices, including scientific vocabulary, for effective expression of meaning. The text uniformly conveys an impression of correctness and scientific accuracy with few, if any, errors. Errors that occur may appear as a consequence of risk-taking in language use or an elaboration upon a science concept or process. The writer consistently: • uses varied sentence formation to create style and tone and to enhance meaning • uses language, including scientific vocabulary, to create style and tone and to enhance meaning • demonstrates correct language usage, punctuation, spelling and capitalization.
Proficient 3	Throughout much of the text there is evidence of the writer using language choices, including scientific vocabulary, for effective expression of meaning. The text generally conveys an impression of correctness. The errors that occur may be of one or two types and occur infrequently. Sometimes errors that occur may appear as a consequence of risk-taking in language use or an elaboration upon a science concept or process. The writer frequently: • uses varied sentence formation to create style and tone and to enhance meaning • uses language choices, including scientific vocabulary, to create style and tone and to enhance meaning • demonstrates correct language usage, punctuation, spelling, and capitalization.
Emergent 2	In portions of the text there is evidence of the writer using language choices, including scientific vocabulary, for effective expression of meaning. Errors of several types may occur, and may be repeated. Errors do not appear to be the result of risk-taking. The writer sometimes: • uses varied sentence formation to create style and tone and to enhance meaning • uses language choices, including scientific vocabulary, to create style and tone and to enhance meaning • demonstrates correct language usage, punctuation, spelling, and capitalization.
Novice 1	In little or none of the text is there evidence of the writer using language choices, including scientific vocabulary, for effective expression of meaning. The text uniformly conveys an overall impression of being error-ridden. Errors do not appear to be the result of risk-taking. The writer rarely or never: • uses varied sentence formation to create style and tone and to enhance meaning • uses language choices to create style and tone and to enhance meaning • demonstrates correct language usage, punctuation, spelling, and capitalization.

Comments	Goals	Actions

Figure 7.29 Language in Use in Science Writing (Holistic Rubric)

Writing Procedures for Conducting Scientific Investigations (Performance List Rubric)

Name _____   Date _____   Course/Class _____

Task/Assignment _____

Performance Criteria	Assessment			
	Points	Self	Teacher	Other(s)
1. Appropriate tools, techniques, and metric units were selected and used effectively for making measurements.				
2. Measuring techniques were practiced and refined before final measurements were recorded.				
3. Careful measurements were taken in order to minimize systematic measurement error.				
4. The set of measurements is recorded in an organized way (list, table, or chart) so that patterns in the data can easily be discerned.				
5. All measurements are clearly labeled with an appropriate magnitude (numerical value) and unit.				
6. Measurements are reported to the correct number of significant figures.				
7. Alternative strategies, techniques, and measuring tools for improving measurements were examined and discussed.				
8. Multiple measurements were repeated to ensure accuracy.				

Comments	Goals	Actions

Figure 7.30   Writing Procedures for Conducting Scientific Investigations (Performance List Rubric)

Measuring Scientifically (Holistic Rubric)

Name _____ Date _____ Course/Class _____

Task/Assignment _____

Expert 4	Appropriate tools, techniques, and metric units were selected and used creatively for making measurements. Measuring techniques were practiced often and refined before final, repeated measurements were recorded, thus eliminating systematic error. All measurements are clearly labeled with an appropriate magnitude (numerical value) and unit. Multiple measurements were repeated to ensure accuracy. Alternative strategies, techniques, and measuring tools for improving measurements were examined and discussed.
Proficient 3	Appropriate tools, techniques, and metric units were selected for making measurements. Measuring techniques were practiced before final, repeated measurements were recorded. Although careful measurements were taken, there is some minor systematic measurement error. The set of measurements is recorded in an organized way. Most measurements are clearly labeled with an appropriate magnitude (numerical value) and unit. Most measurements are reported to the correct number of significant figures.
Emergent 2	Some of the tools, techniques, and metric units are not appropriate for the task. There is no evidence that measuring techniques were practiced before final measurements were recorded. Measurements were taken, but there are major systematic measurement errors. The set of measurements is recorded in a somewhat disorganized way. Some measurements are clearly labeled with an appropriate magnitude (numerical value) and unit. Some measurements are reported to the correct number of significant figures. It is unclear whether the same measurements were repeated.
Novice 1	Most of the tools, techniques, and metric units are not appropriate for the task. There is no evidence that measuring techniques were practiced before final measurements were recorded. Few, if any, measurements were taken and there are major systematic measurement errors. The set of measurements is recorded in an extremely disorganized way. Measurements are not clearly labeled with an appropriate magnitude (numerical value) and unit. No measurements are reported to the correct number of significant figures. Measurements were not repeated.

Comments	Goals	Actions

Figure 7.31   Measuring Scientifically (Holistic Rubric)

Measuring Scientifically (Analytic Rubric)

Name _____ Date _____ Course/Class _____

Task/Assignment _____

	Tools, Techniques, and Units	Measuring Techniques	Reporting Measurements	Alternative Strategies
Expert (4)	Appropriate tools, techniques, and metric units were selected and used creatively for making measurements. ☐	Measuring techniques were practiced often and refined before final measurements were recorded. Careful measurements were taken in order to minimize systematic measurement error. ☐	The set of measurements is recorded in an organized way (list, table, or chart) so that patterns in the data can easily be discerned. All measurements are clearly labeled with an appropriate magnitude (numerical value) and unit. Measurements are reported to the correct number of significant figures. Multiple measurements were repeated to ensure accuracy. ☐	Alternative strategies, techniques, and measuring tools for improving measurements were examined and discussed. ☐
Proficient (3)	Appropriate tools, techniques, and metric units were selected for making measurements. ☐	Measuring techniques were practiced before final measurements were recorded. Although careful measurements were taken, there is some minor systematic measurement error. ☐	The set of measurements is recorded in an organized way (list, table, or chart). Most measurements are clearly labeled with an appropriate magnitude (numerical value) and unit. Most measurements are reported to the correct number of significant figures. Some measurements were repeated to ensure accuracy. ☐	Alternative strategies, techniques, and measuring tools for improving measurements were not examined and discussed. ☐

Figure 7.32 Measuring Scientifically (Analytic Rubric)

(Continued)

Measuring Scientifically (continued)

	Tools, Techniques, and Units	Measuring Techniques	Reporting Measurements	Alternative Strategies
Emergent (2)	Some of the tools, techniques, and metric units are not appropriate for the task. ❏	There is no evidence that measuring techniques were practiced before final measurements were recorded. Measurements were taken, but there are major systematic measurement errors. ❏	The set of measurements is recorded in a somewhat disorganized way. Some measurements are clearly labeled with an appropriate magnitude (numerical value) and unit. Some measurements are reported to the correct number of significant figures. It is unclear whether the same measurements were repeated. ❏	Alternative strategies, techniques, and measuring tools for improving measurements were not examined and discussed. ❏
Novice (1)	Most of the tools, techniques, and metric units are not appropriate for the task. ❏	There is no evidence that measuring techniques were practiced before final measurements were recorded. ❏	Few, if any, measurements were taken and there are major systematic measurement errors. The set of measurements is recorded in an extremely disorganized way. Measurements are not clearly labeled with an appropriate magnitude (numerical value) and unit. No measurements are reported to the correct number of significant figures. Measurements were not repeated. ❏	Alternative strategies, techniques, and measuring tools for improving measurements were not examined and discussed. ❏

140

Observing and Inferring in Science (Performance List Rubric)

Name _____ Date _____ Course/Class _____

Task/Assignment _____

	Assessment			
Performance Criteria	**Points**	**Self**	**Teacher**	**Other(s)**
1. Observations are based upon what was actually observed and not based upon prior knowledge, personal opinion, observer bias, or inferences.				
2. Appropriate tools and materials were selected, evaluated, and then used to make the final observations.				
3. Appropriate metric measurements are used to describe quantitative observations.				
4. Observations are quantitatively and/or qualitatively accurate.				
5. Both magnitude and units are recorded for quantitative data.				
6. Observations are interpreted by comparing and contrasting objects or events.				
7. Inferences are explained and justified based upon background research, investigative data, and /or the observer's prior knowledge.				
8. Inferences fall within a range of acceptance (reasonableness) as based upon all the observations, data, and the observer's prior experience.				

Comments	Goals	Actions

Figure 7.33   Observing and Inferring in Science (Performance List Rubric)

Observing and Inferring in Science (Holistic Rubric)

Name _____ Date _____ Course/Class _____

Task/Assignment _____

Expert 4	Observations are based upon what was actually observed and not inferred. Appropriate tools and materials were selected, evaluated, and then used to make the final observations. Appropriate and accurate metric measurements are used to describe quantitative observations, with both magnitude and units being recorded. Inferences are explained and justified based upon background research, investigative data, and /or the observer's prior knowledge. Inferences fall within a range of acceptance (reasonableness) as based upon all the observations, data, and the observer's prior experience.
Proficient 3	Most observations are based upon what was actually observed and not inferred. Tools and materials were selected, evaluated, and then used to make the final observations. Metric measurements, with minor errors, are used to describe quantitative observations, with both magnitude and units being recorded. Most inferences are explained and justified based upon background research, investigative data, and /or the observer's prior knowledge. Most inferences fall within a range of acceptance (reasonableness) as based upon all the observations, data, and the observer's prior experience.
Emergent 2	Some observations are based upon what was actually observed and not inferred. Some tools and materials were selected, evaluated, and then used to make the final observations. There are major measurement errors. Some inferences are explained and justified based upon background research, investigative data, and /or the observer's prior knowledge. Some inferences fall within a range of acceptance (reasonableness) as based upon all the observations, data, and the observer's prior experience.
Novice 1	Few, if any, observations are based upon what was actually observed and not inferred. Inappropriate tools and materials were selected, evaluated, and then used to make the final observations. There are major measurement errors. Few, if any, inferences are explained and justified based upon background research, investigative data, and/or the observer's prior knowledge. Few if any inferences fall within a range of acceptance (reasonableness) as based upon all the observations, data, and the observer's prior experience.

Comments	Goals	Actions

Figure 7.34 Observing and Inferring in Science (Holistic Rubric)

Observing and Inferring in Science (Analytic Rubric)

Name _____ Date _____ Course/Class _____

Task/Assignment

	Basis for Observations	Tools and Materials	Describing Observations	Inferences
Expert (4)	Observations are based upon what was actually observed and not based upon prior knowledge, personal opinion, observer bias, or inferences. ❑	Appropriate tools and materials were selected, evaluated, and then used to make the final observations. ❑	Appropriate metric measurements are used to describe quantitative observations. Observations are quantitatively and/or qualitatively accurate. Both magnitude and units are recorded for quantitative data. Observations are interpreted by comparing and contrasting objects or events. ❑	Inferences are explained and justified based upon background research, investigative data, and/or the observer's prior knowledge. Inferences are supported by and directly linked to all the observations and the data. Inferences fall within a range of acceptance (reasonableness) as based upon all the observations, data, and the observer's prior experience. ❑
Proficient (3)	For the most part, observations are based upon what was actually observed and not based upon prior knowledge, personal opinion, observer bias, or inferences. ❑	Appropriate tools and materials were selected and then used to make the final observations. ❑	Appropriate metric measurements are used to describe quantitative observations. Observations are quantitatively and/or qualitatively accurate, with few minor errors. Both magnitude and units are recorded for quantitative data. There is some attempt to interpret the observations by comparing and contrasting objects or events. ❑	Inferences are explained with some justification provided, based upon background research, investigative data, and /or the observer's prior knowledge. Inferences are supported by and directly linked to most of the observations and the data. Inferences fall within a range of acceptance (reasonableness) as based upon all the observations, data, and the observer's prior experience. ❑

Figure 7.35 Observing and Inferring in Science (Analytic Rubric)

(Continued)

Observing and Inferring in Science (continued)

	Basis for Observations	Tools and Materials	Describing Observations	Inferences
Emergent (2)	Some of the observations are not based upon what was actually observed, but instead are based upon prior knowledge, personal opinion, observer bias, or inferences. ❐	There is some question as to the appropriateness of the tools and materials selected to make the final observations. There is no evidence the appropriateness of the tools or materials was evaluated. ❐	Appropriate metric measurements are used to describe quantitative observations. Major errors of accuracy exist within the observations. Some magnitudes and units are missing for quantitative data. There is no attempt to interpret the observations by comparing and contrasting objects or events. ❐	Inferences are explained without justification provided. Some inferences are supported by and directly linked to most of the observations and the data. Some inferences fall outside the range of acceptance (reasonableness). ❐
Novice (1)	Most of the observations are not based upon what was actually observed, but instead are based upon prior knowledge, personal opinion, observer bias, or inferences. ❐	Many questions exist as to the appropriateness of the tools and materials selected to make the final observations. There is no evidence the appropriateness of the tools or materials was ever evaluated. ❐	Some metric measurements are used to describe quantitative observations, but many errors exist. Major errors of accuracy exist within the observations. Some magnitudes and most units are missing for quantitative data. There is no attempt to interpret the observations by comparing and contrasting objects or events. ❐	Little attempt is made to explain inferences. Explanations are unsupported by the observations and the data. Most inferences fall outside the range of acceptance (reasonableness). ❐

144

Oral Presentation in Science (Performance List Rubric)

Name _____ Date _____ Course/Class _____

Task/Assignment _____

Performance Criteria	Assessment			
	Points	Self	Teacher	Other(s)
Content and Organization				
1. The purpose of the presentation (informing, persuading or both), the subject, and any position taken by the presenter are clearly defined at the outset.				
2. The presentation is made in an interesting, logical sequence – an introduction, an organized body, and a clear closure – that the audience can follow.				
3. The introduction has a strong purpose statement that serves to captivate the audience and narrow the topic.				
4. An abundance of accurate supporting scientific concepts, facts, figures, statistics, scenarios, stories, and analogies are used to support the key points and ideas.				
5. The vocabulary is appropriate to both the science content and the audience.				

Figure 7.36   Oral Presentation in Science (Performance List Rubric)

(Continued)

Oral Presentation in Science (continued)

Performance Criteria	Assessment			
	Points	**Self**	**Teacher**	**Other(s)**
Optional				
6. Interesting and colorful audiovisual aids or multimedia materials are interwoven to explain and reinforce the screen text and presentation.				
7. The topic is developed completely and thoroughly.				
Presentation				
8. The speaker maintains a proper volume, clear elocution, steady rate, effective inflections, and enthusiasm throughout the presentation.				
9. Humor is used positively and in good taste, with consideration given to the composition of the audience.				
10. Stories and motivational scenarios are used appropriately.				
11. Body language such as eye contact, posture, gestures, and body movements are appropriate and are used to create effect.				
12. Delivery is well paced, flows naturally, has good transitions, and is coherent.				
13. The speaker is relaxed, self-confident, and appropriately dressed for purpose or audience.				

Oral Presentation in Science (continued)

Performance Criteria	Assessment			
	Points	Self	Teacher	Other(s)
Audience 14. The audience's attention is maintained by involving them in the presentation.				
15. Information needed by audience to fully understand the presentation is provided.				
16. The speaker gives the audience time to think, reflect, and ask questions about points made in the presentation.				
17. The speaker answers all questions with clear explanations and further elaborations.				
18. The topic and the length of the presentation are appropriate for the audience and within the allotted time limits.				

Comments	Goals	Actions

Oral Presentation in Science (Analytic Rubric)

Name _____ Date _____ Course/Class _____

Task/Assignment _____

	Purpose	Content	
		Scientific Accuracy	**Supporting Data**
Expert (4)	The purpose of the presentation – either informing, persuading or both – the subject, and any position taken by the presenter are clearly defined at the outset. ☐	Contains relevant and complete scientific concepts, facts, figures, statistics, scenarios, and/or stories. ☐	Key points or ideas are completely and accurately supported with data and/or evidence. ☐
Proficient (3)	Some success in defining purpose, subject, and position; information and data presented are generally consistent with purpose. ☐	Contains some relevant scientific concepts, facts, figures, statistics, scenarios, and/or stories. ☐	Key points or ideas are supported with some data and/or evidence. ☐
Emergent (2)	Attempts to define purpose, subject, and position with some success; but presents contradictory information and data. ☐	Contains weak scientific concepts, facts, figures, statistics, scenarios, and/or stories, which do not support the subject. ☐	Key points or ideas are supported with very little data and/or evidence. ☐
Novice (1)	Subject, purpose, and position defined with no success; presents non-related information and data. ☐	Contains very weak support of subject through scientific concepts, facts, figures, statistics, scenarios, and/or stories. ☐	Totally insufficient support for key points or ideas. ☐

Figure 7.37 Oral Presentation in Science (Analytic Rubric)

Oral Presentation in Science (continued)

	Organization		
	Introduction	**Body**	**Closure**
Expert (4)	Introduction has strong purpose statement that serves to captivate the audience and narrow the topic. ❏	Topic is researched, narrowed, and organized with only relevant and supportive information being presented. ❏	Audience responds enthusiastically and positively to presentation, major ideas have been summarized, audience leaves with a full understanding of presenter's position. ❏
Proficient (3)	Introductory statement informs audience of general purpose of presentation. ❏	Topic needs some further research, narrowing, and organization. Some extraneous information is included. ❏	Audience responds somewhat positively to presentation, major ideas have been summarized, audience leaves with an understanding of presenter's position. ❏
Emergent (2)	Introduction of subject fails to make audience aware of the purpose of presentation. ❏	Topic is insufficiently researched, too broad, and/or disorganized. Much extraneous information is included. ❏	Audience responds less than positively to presentation, major ideas have been summarized, but in an unclear fashion, the audience leaves with some understanding of presenter's position. ❏
Novice (1)	No introductory statement or an introductory statement that confuses audience. ❏	Topic is general, vague, and/or disorganized. ❏	Major ideas left out and audience leaves with no understanding of presenter's position. ❏

(Continued)

Oral Presentation in Science (continued)

	Presentation Aids	Poise/ Appearance	Body Language	Delivery: Voice	Pacing	Eye Contact
Expert (4)	Are clear, appropriate, not over used, and beneficial to the speech. ❑	Relaxed, self-confident, and appropriately dressed for purpose or audience. ❑	Natural movement and descriptive gestures that display energy, create mood, and help audience visualize. ❑	Fluctuation in volume and inflection help to maintain audience interest and emphasize key points. ❑	Good use of pause, giving sentence drama; length matches allotted time. ❑	Maintains direct eye contact with all parts of audience. ❑
Proficient (3)	Are used and add some clarity and dimension to speech. ❑	Quick recovery from minor mistakes; appropriately dressed. ❑	Movements and gestures generally enhance delivery. ❑	Satisfactory variation of volume and inflection. ❑	Pattern of delivery generally successful; slight mismatch between length and allotted time. ❑	Fairly consistent use of direct eye contact with audience. ❑
Emergent (2)	Attempted, but unclear, inappropriate or over-used. ❑	Some tension or indifference apparent and possible inappropriate dress for purpose or audience. ❑	Insufficient movement and/or awkward gestures. ❑	Uneven volume with little or no inflection. ❑	Uneven or inappropriate patterns of delivery and/or length that do not fit within allotted time. ❑	Occasional but unsustained eye contact with audience. ❑
Novice (1)	None attempted. ❑	Nervous tension obvious and/or inappropriately dressed for purpose or audience. ❑	No movement or descriptive gestures. ❑	Low volume and/or monotonous tone cause audience to disengage. ❑	Delivery is either too fast or too slow and/or does not fit within allotted time. ❑	No effort to maintain eye contact with audience. ❑

Oral Presentation in Science (continued)

	Audience			
	Ability to Inform	**Ability to Persuade**	**Use of Humor/Stories**	**Language Usage**
Expert (4)	Significantly increases audience understanding and knowledge of topic. Information needed by the audience to fully understand the presentation is skillfully developed. ❏	Effectively convinces an audience to recognize the validity of point of view. ❏	Uses humor appropriately to make significant points about the topic. Stories and motivational scenarios are used appropriately. ❏	Language chosen is very descriptive and appropriate for audience. Vocabulary and sentence structure are on an intellectual level comparable to audience. ❏
Proficient (3)	Raises audience understanding and awareness of most points. ❏	Point of view is mostly clear, but not fully developed. ❏	Achieves moderate success in using humor, stories, and motivational scenarios. ❏	Language chosen, for the most part, is appropriate for audience. Vocabulary and sentence structure are on an appropriate intellectual level for the audience. ❏
Emergent (2)	Raises audience understanding and knowledge of some points. ❏	Point of view is somewhat clear, but lacks development or support. ❏	Humor attempted but inconsistent, weak, or inappropriate. ❏	Language chosen is appropriate for most of the audience. Vocabulary and sentence structure lack appropriateness. ❏
Novice (1)	Fails to raise audience understanding and knowledge of topic. ❏	Fails to effectively convince the audience of point of view. ❏	No use of humor or humor used inappropriately. ❏	Language chosen is inappropriate for most of the audience. Vocabulary and sentence structure miss the target audience. ❏

Pamphlet (Performance List Rubric)

Name _____ Date _____ Course/Class _____

Task/Assignment _____

Performance Criteria	Assessment			
	Points	Self	Teacher	Other(s)
1. Scientific content is accurate and supports the major sections of the pamphlet.				
2. The format used to lay out the pamphlet is effective for the intended audience.				
3. The pamphlet is creative and interesting.				
4. The writing in the pamphlet is objective, clear and concise, has clear sentence structure and uses descriptive rather than figurative language.				
5. The pamphlet has all the following presentation elements: words and visuals are easy to see, titles and headings are easy to distinguish, and colors and patterns in the pamphlet are pleasing.				
6. Diagrams, pictures, and other graphics are of quality and add to the overall effectiveness of the pamphlet.				
7. References are included and are correctly cited.				
8. There are no errors in the mechanics (spelling and grammar).				

Comments	Goals	Actions

Figure 7.38 Pamphlet (Performance List Rubric)

Portfolio (Performance List Rubric)

Name _____ Date _____ Course/Class _____

Task/Assignment _____

Performance Criteria	Assessment			
	Points	**Self**	**Teacher**	**Other(s)**
The Collection of Artifacts				
1. Processes of science: the collection of artifacts shows understanding of scientific processes including measuring, collecting data, organizing data, and designing and conducting experiments, etc.				
2. Concepts of Science: the collection of artifacts shows an in-depth understanding of selected science concepts.				
3. Nature of Science: the artifacts selected reflect an understanding of the Nature of Science.				
4. Connections between science and technology: the collection of artifacts demonstrates the use of technology to answer questions and solve problems.				
Optional				
5. Connections between science and mathematics: the collection of artifacts demonstrates the use of mathematics to answer questions and solve problems.				
6. Connections between science and Reading/English Language Arts: the collection of artifacts demonstrates the use of Reading/English Language Arts to communicate the answers to questions and solve problems.				
7. Science products/performances: the collection of artifacts includes things such as science fair projects, lab reports, student service learning projects, experiments, brief and extended constructed responses to open-ended questions, scientific drawings, models, and Internet searches.				

Figure 7.39   Portfolio (Performance List Rubric)

(Continued)

Portfolio (continued)

Performance Criteria	Assessment			
	Points	Self	Teacher	Other(s)
The Overall Portfolio				
8. The portfolio has a clearly labeled cover including the subject, the student's and the teacher's name, and any illustrations that add to explaining the contents.				
9. The portfolio contains a collection of artifacts, a table of contents, and a self-reflection narrative for each category.				
10. The table of contents for each category is clear.				
11. The self-reflection narrative in each category addresses why specific artifacts were chosen for inclusion.				
12. The self-reflection narrative in each category addresses strengths and weaknesses of artifacts chosen for that category.				
13. The self-reflection narrative in each category tells about new skills and content learned for that category.				

Comments	Goals	Actions

Poster (Performance List Rubric)

Name _____   Date _____   Course/Class _____

Task/Assignment _____

Performance Criteria	Assessment			
	Points	**Self**	**Teacher**	**Other(s)**
1. The poster contains a title that clearly reflects the topic or theme.				
2. The poster contains relevant and accurate information about the topic or theme.				
3. The format of the poster is appropriate to the content, purpose, and audience for which it is designed.				
4. Graphic elements, such as pictures, photographs, charts, tables, scientific drawings, diagrams, graphs, etc., add to the overall effectiveness of the poster				
5. There is a coherent, flowing organization to the poster with the various elements (text, graphics, etc.) working well together.				
6. The poster is aesthetically pleasing, with effective use of space, color, texture, and shape.				
7. The poster is skillfully designed and crafted using appropriate graphic design tools.				
8. The poster effectively communicates its theme in convincing fashion to the intended audience.				
9. The poster is creative and draws attention.				
10. Language chosen for the poster is captivating, persuasive, informative, accurate, and concise.				

Comments	Goals	Actions

Figure 7.40   Poster (Performance List Rubric)

PowerPoint Presentation (Performance List Rubric)

Name _____ Date _____ Course/Class _____

Task/Assignment _____

	Assessment			
Performance Criteria	**Points**	**Self**	**Teacher**	**Other(s)**
1. The topic has been extensively and accurately researched using varied, current, and appropriate information sources.				
2. A storyboard, consisting of logically and sequentially numbered slides, has been developed which includes a thumbnail sketch of each slide. The sketch contains the title, text, background color, graphics, fonts, and hyperlinks with URLs for each slide.				
3. The introduction is captivating and engages the audience.				
4. The content is written clearly, accurately, and completely in a logical fashion.				
5. The fonts are easy to read and point size varies appropriately for headings and text. The use of italics, bold, and underline contributes to the readability of the text.				
6. The background, colors, and graphics enhance the readability and aesthetics of the text.				
7. The graphics, animation, and sound enhance the overall presentation and contribute to an understanding of the concepts, ideas, and themes.				

Figure 7.41 PowerPoint Presentation (Performance List Rubric)

PowerPoint Presentation (continued)

Performance Criteria	Assessment			
	Points	Self	Teacher	Other(s)
8. Graphics are of proper size and resolution and are strategically placed to enhance comprehension.				
9. The layout is aesthetically pleasing and contributes to an understanding of the underlying concepts, ideas, and themes.				
10. Transitions act as visually interesting pauses between screens and convey meaning, such as a topic shift or shift in level of detail.				
11. Language chosen for the presentation is captivating, persuasive, informative, accurate, and concise.				
12. The text is free of spelling, punctuation, capitalization, and grammatical errors.				
Optional Video clips that are added to the presentation enhance comprehension of main concepts.				
Images used in the presentation are from three or more sources (scan, CD-ROM, videotape, Web, Photoshop, digital camera, etc.)				

Comments	Goals	Actions

Science Fair Display (Performance List Rubric)

Name _____ Date _____ Course/Class _____

Task/Assignment _____

Performance Criteria	Assessment			
	Points	Self	Teacher	Other(s)

Overall Design

1. Overall appearance of backboard is organized, logical, and legible.

2. There is a symmetry, flow, and completeness to the backboard. It does not appear as a collection of facts and figures.

3. No more than 3 colors were used when constructing the backboard.

4. Graphics that are legible, attractive, and add to the interpretation of the backboard are included.

5. All necessary parts are labeled and placed in a logical order. (Question, Hypothesis, Materials, Procedures, Data, Summary of Data, and Conclusion)

6. The backboard has a title that clearly describes the dependent and independent variables and the results.

7. All of the words on the backboard are spelled correctly.

8. Language used on the backboard is concise, purposeful, and scientifically accurate.

9. A research paper containing a bibliography on the topic is placed in front of the backboard.

10. The journal/log containing notes, observations, and data collected during the experiment is also placed in front of the backboard.

Question and Hypothesis

11. The question and hypothesis are stated in such a way as to guide the development of an investigation.

12. The hypothesis is stated in an "If-and-then" format.

Figure 7.42   Science Fair Display (Performance List Rubric)

Science Fair Display (continued)

	Assessment			
Performance Criteria	**Points**	**Self**	**Teacher**	**Other(s)**

Procedures

13. The procedures are written in a clear, concise, step-by-step fashion and could be followed by another person wanting to duplicate the experiment.

14. Both the independent and dependent variables are identified within the procedures.

15. A description of how the dependent variable is measured is contained within the procedures.

16. Controls are identified within the procedures.

17. Repeated testing (trials) of the independent variable is described within the procedures.

Data Organization and Analysis

18. Data are organized into tables, charts, or other graphic display techniques.

19. Appropriate statistics and graphic analyses are applied to develop inferences from the data.

20. Data are analyzed statistically and graphically for patterns and trends.

21. All graphs and statistical analysis are conducted correctly.

Inferences and Conclusions

22. The conclusions state whether the hypothesis was accepted or rejected and why.

23. The conclusions state only the major findings of the experiment and are supported directly by the data.

24. At least one question for further study is included in the conclusions.

Comments	Goals	Actions

Scientific Drawing (Performance List Rubric)

Name _____   Date _____   Course/Class _____

Task/Assignment _____

Performance Criteria	Assessment			
	Points	**Self**	**Teacher**	**Other(s)**
1. The drawing(s) realistically and effectively depict(s) the object(s).				
2. The drawing includes only those features that were actually observed and not inferred.				
3. Many relevant details are included: size (with metric measurements), colors, textures, shapes, and relationships to surroundings.				
4. Multiple perspectives are drawn to provide the viewer with a complete picture of the structures under study.				
5. A descriptive and accurate title is provided for the drawing(s).				
6. All the parts of the scientific drawing are clearly and accurately labeled.				

Figure 7.43   Scientific Drawing (Performance List Rubric)

Scientific Drawing (continued)

Performance Criteria	Assessment			
	Points	**Self**	**Teacher**	**Other(s)**
7. A detailed, written explanation of what the scientific drawing is intended to show is included.				
8. A key or legend, if needed to explain the drawing(s), is provided.				
9. The scientific drawing(s) is/are of an appropriate size and scale for details to be easily recognized.				
10. A very precise scale and proportion is used consistently. The scale is stated and uses the metric system when possible.				
11. The scientific content is accurately represented and is appropriate for the drawing.				

Comments	Goals	Actions

Scientific Drawing (Holistic Rubric)

Name _____ Date _____ Course/Class _____

Task/Assignment _____

Expert 4	The drawing(s) realistically and effectively depict(s) the object(s). Multiple perspectives are provided to enhance understanding. A descriptive and accurate title is provided and all the parts of the drawing are clearly labeled. A detailed written explanation of what the scientific drawing is intended to show is included, along with a key or legend to further explain the drawing(s). The drawing(s) is/are of an appropriate size and consistent metric scale for details to be easily recognized. The scientific content is accurately represented and is appropriate for the drawing.
Proficient 3	The drawing(s) depict(s) the object(s). Many details are included. A descriptive and accurate title is provided and most parts of the drawing are clearly and neatly labeled. A sketchy written explanation of what the scientific drawing is intended to show is included. The drawing(s) is/are of an appropriate size and scale for details to be easily recognized. The scientific content is accurately represented and is appropriate for the drawing.
Emergent 2	The drawing(s) reasonably depict(s) the object(s). The drawing(s) is a reasonable rendition of the object(s), but may include features that were not actually observed. Some details are included. Only one perspective of the object(s) is provided. A title is provided for the drawing(s). Some parts of the scientific drawing are labeled. Labeling lacks neatness, legibility, and attractiveness. A sketchy written explanation of what the scientific drawing is intended to show is included. The drawing(s) is/are inappropriately sized and scaled. The scientific content contains some inaccuracies.
Novice 1	The drawing(s) is/are clearly lacking in realism, accuracy, and detail. It is difficult to tell what the drawing(s) represent(s). Scale and proportion are clearly lacking. Metric measurements are missing. Few distinguishing forms, structures, and details are labeled. Labeling is not consistently neat, legible, and attractive. No attempt is made to provide a title of the drawing(s). The scientific content contains many inaccuracies.

Comments	Goals	Actions

Figure 7.44   Scientific Drawing (Holistic Rubric)

Scientific Drawing (Analytic Rubric)

Name _____   Date _____   Course/Class _____

Task/Assignment _____

	Accuracy and Realism	Scale and Proportion	Labeling	Titles and Accompanying Text
Expert (4)	The drawing(s) realistically and effectively depict(s) the object(s). Amazing detail is provided for size, color, texture, and shape. Multiple perspectives are provided to clearly distinguish form, structures, and dimensions. The scientific content is accurately represented and is appropriate for the drawing. ❑	A very precise scale and proportion is provided using metric measurements. The scale and proportion are appropriate for showing details. The scale is stated either in the drawing itself or the accompanying key or legend. The relationship between the object and its environment is shown. ❑	All distinguishing forms, structures, and details are clearly labeled. Labeling is consistently neat, legible, and attractive in appearance. ❑	A descriptive and accurate title of the drawing(s) is provided. A detailed, interpretative, written explanation of what the drawing(s) is/are intended to show is provided. ❑
Proficient (3)	The drawing(s) depict(s) the object(s). Amazing detail is provided for size, color, texture, and shape. Multiple perspectives are missing. The scientific content is accurately represented and is appropriate for the drawing. ❑	A very precise scale and proportion is provided using metric measurements. The scale and proportion are appropriate for showing details. The legend or key of the scale is missing. The relationship between the object and its environment is shown. ❑	All distinguishing forms, structures, and details are clearly labeled. Labeling is not consistently neat, legible, and attractive in appearance. ❑	A title of the drawing(s) is provided. A written explanation of what the drawing(s) is/are intended to show is provided. However, in both cases, details and clarity are lacking. ❑

Figure 7.45 Scientific Drawing (Analytic Rubric)

(Continued)

163

Scientific Drawing (continued)

	Accuracy and Realism	Scale and Proportion	Labeling	Titles and Accompanying Text
Emergent (2)	The drawing(s) reasonably depict(s) the object(s). Many details are provided. Multiple perspectives are missing. The scientific content contains some inaccuracies. ❑	A rather imprecise scale and proportion is provided. Metric measurements are missing. The scale and proportion are appropriate for showing details. The legend or key of the scale is missing. The relationship between the object and its environment is not shown. ❑	Most distinguishing forms, structures, and details are clearly labeled. Labeling is not consistently neat, legible, and attractive in appearance. ❑	An attempt is made to provide a title of the drawing(s). A written explanation of what the drawing(s) is/are intended to show is not provided. ❑
Novice (1)	The drawings are clearly lacking in how realistic the object(s) is/are drawn. It is difficult to tell from the drawing(s) what the object is. Few details are provided. Multiple perspectives are missing. The scientific content contains many inaccuracies. ❑	Scale and proportion are clearly lacking. Metric measurements are missing. The scale and proportion are not appropriate for showing details. The legend or key of the scale is missing. The relationship between the object and its environment is not shown. ❑	Few distinguishing forms, structures, and details are clearly labeled. Labeling is not consistently neat, legible, and attractive in appearance. ❑	No attempt is made to provide a title of the drawing(s). A written explanation of what the drawing(s) is/are intended to show is not provided. ❑

Scientific Investigation (Performance List Rubric)

Name _____ Date _____ Course/Class _____

Task/Assignment _____

Performance Criteria	Assessment			
	Points	Self	Teacher	Other(s)
1. The question has been developed in such a way that it can be answered by conducting an experiment(s).				
2. The hypothesis has been developed directly from the question and is expressed in an "If-and-then" statement(s).				
3. The design of the experiment tests the hypothesis.				
4. The procedures follow a logical step-by-step order and include a list of all necessary materials.				
5. The procedures are written clearly enough so that another person could repeat this experiment.				
6. The procedures show that repeated trials were done.				
7. Both the dependent and independent variables have been identified.				
8. The experimental design uses proper controls and tests for the effects of only one independent variable at a time.				
9. The conclusions of the experiment are written in clear and complete statements, and are supported by the inferences.				
10. Language is used purposefully and written in complete sentences when writing the question, hypothesis, procedures, results, and conclusions.				

Comments	Goals	Actions

Figure 7.46   Scientific Investigation (Performance List Rubric)

Scientific Investigation (Holistic Rubric)

Name _____ Date _____ Course/Class _____

Task/Assignment _____

Expert 4	The question has been developed in such a way that it can be answered by conducting an experiment and reflects background research and previous observations. The hypothesis has been developed directly from the question and is expertly expressed in an "If-and-then" statement(s). The procedures are detailed, complete, follow a logical step-by-step order, and include a list of all necessary materials. The experimental design uses proper controls and tests for the effects of only one independent variable at a time. The collected data are organized and displayed in appropriate graphic formats. The data are manipulated through the use of appropriate statistical methods. The conclusions of the experiment are written in clear and complete statements and are supported by the data. Language used is appropriate, purposeful, and written in complete sentences. Scientific content and terminology are accurate.
Proficient 3	The question provides general guidance to the design of an experiment. The hypothesis has been developed from the question and is expressed in an "If-and-then" statement(s). The procedures are complete, follow a somewhat logical step-by-step order, and include a list of materials. The experimental design uses proper controls and tests for the effects of only one independent variable at a time. The collected data are organized, displayed, and manipulated through the use of appropriate statistical methods. The conclusions of the experiment are written in clear and complete statements and are mostly supported by the data. Language used is appropriate and purposeful. Scientific content and terminology may contain minor errors.
Emergent 2	The question provides some guidance to the design of an experiment. The hypothesis is loosely connected to the question and there is an attempt to express it in an "If-and-then" statement(s). The procedures are incomplete and follow a somewhat illogical step-by-step order. The experimental design does not completely identify nor control variables. The collected data are disorganized and there is limited manipulation through the use of appropriate statistical methods. The conclusions of the experiment are loosely supported by the data. Much of the language used is inappropriate. Scientific content and terminology contains major errors.
Novice 1	The question is ill defined and gives little to no direction for developing an experiment. The hypothesis bears little to no connection to the question. The design of the experiment is unclear. The procedures are confusing and difficult to follow. Variables have not been clearly identified or controlled. The conclusions of the experiment are vague, not written in clear and complete statements, and are not supported by the data.

Comments	Goals	Actions

Figure 7.47 Scientific Investigation (Holistic Rubric)

Scientific Investigation (Analytic Rubric)

	Question and Hypothesis	Experiment Design	Conclusion	Language Usage and Scientific Content
Expert **(4)**	The question has been written in such a way that it can be answered by conducting an experiment and reflects both background research and previous observations. The hypothesis has been developed directly from the question and is expertly expressed in "If-and-then" statements. ❑	The design of the experiment tests the hypothesis and reflects creative use of both materials and equipment. The procedures are detailed, complete, follow a logical step-by-step order, and include a list of all necessary materials. The procedures are expertly written in such a way that another person could easily repeat the experiment. The procedures show that repeated trials were done. Both the dependent and independent variables have been identified. The experimental design uses proper controls and tests for the effects of only one independent variable at a time. The collected data are organized and displayed in appropriate graphic formats. The data are manipulated through the use of appropriate statistical methods. ❑	The conclusions of the experiment are written in clear and complete statements and are supported by the data. The hypothesis is accepted or rejected with data being cited and the question to the experiment is answered within the conclusions. Potential questions for future research that arose during the experiment are addressed in the conclusions. ❑	Language is used purposefully and written in complete sentences when writing the question, hypothesis, procedures, results, and conclusions. The language is appropriate for the intended audience. The scientific content and terminology are accurate. ❑
Proficient **(3)**	The question has been written in such a way that it can be answered by conducting an	The design of the experiment tests the hypothesis. The procedures follow a logical step-by-step order and include a list of all necessary materials. The	The conclusions of the experiment are written in clear and complete statements and are supported	Language is used purposefully and written in complete sentences when writing the question, hypothesis, procedures, results, and

Figure 7.48 Scientific Investigation (Analytic Rubric)

(Continued)

Scientific Investigation (continued)

	Question and Hypothesis	Experiment Design	Conclusion	Language Usage and Scientific Content
	experiment. The hypothesis has been developed directly from the question and is expressed in "If-and-then" statements. ❑	procedures are written clearly enough so that another person could repeat the experiment. The procedures show that repeated trials were done. Both the dependent and independent variables have been identified. The experimental design uses proper controls and tests for the effects of only one independent variable at a time. ❑	by the data. The hypothesis is accepted or rejected and the question to the experiment is answered within the conclusions. Potential questions for future research that arose during the experiment are addressed in the conclusions. ❑	conclusions. The language is appropriate for the intended audience. The scientific content and terminology are generally accurate. ❑
Emergent (2)	The question provides some direction for conducting an experiment. The hypothesis is loosely linked to the question and may not be expressed in "If-and-then" statements. ❑	The design of the experiment does not fully test the hypothesis. Only parts of the procedures follow a logical step-by-step order. The procedures lack clarity. It is difficult to tell if repeated trials were done. Variables and controls are ill defined. ❑	The conclusions of the experiment are written in unclear and incomplete statements, with some support from the data. The hypothesis is neither accepted nor rejected and the question to the experiment is partially answered within the conclusions. ❑	Language is not always used purposefully nor written in complete sentences when writing the question, hypothesis, procedures, results, and conclusions. The language is partially appropriate for the intended audience. The scientific content and terminology contain some inaccuracies. ❑
Novice (1)	The question is ill defined and gives little to no direction for developing experiments. The hypothesis reflects little to no connection to the question. ❑	The design of the experiment is unclear. The procedures are confusing and difficult to follow. Repeated trials were not done. Variables have not been clearly identified, nor controlled. ❑	The conclusions of the experiment are vague and are not written in clear and complete statements, and are not supported by the data. ❑	Language is not used purposefully and is written in incomplete sentences when writing the question, hypothesis, procedures, results, and conclusions. The language is not appropriate for the intended audience. The scientific content and terminology contain many inaccuracies. ❑

Scientific Models (Physical) (Performance List Rubric)

Name _____ Date _____ Course/Class _____

Task/Assignment _____

Performance Criteria	Assessment			
	Points	Self	Teacher	Other(s)
1. Written plans/diagrams of the model were developed, reviewed, and revised before constructing the actual physical model.				
2. The written plan/diagram is drawn to scale, with the dimensions and parts labeled.				
3. The written plan uses metric measurements throughout.				
4. A written explanation accompanies the physical model to describe what the model represents.				
5. Any safety issues regarding the model are described in the written explanation.				
6. The physical model is an accurate replication of the written plan/diagram.				
7. Limitations and strengths of the model are included in the written explanation.				
8. The model is a scientifically accurate reproduction of the real item.				
9. The model has been constructed with care and attention to details.				
10. Features that would enhance the model have been added – color, texture, labels, etc.				

Comments	Goals	Actions

Figure 7.49 Scientific Models (Physical) (Performance List Rubric)

Summarizing Scientific Articles (Performance List Rubric)

Name _____ Date _____ Course/Class _____

Task/Assignment _____

Performance Criteria	Points	Self	Teacher	Other(s)
	Assessment			
1. The summary succinctly describes the most important findings of the article.				
2. Each important finding is supported with relevant information from the article.				
3. The summary accurately interprets science facts, concepts, and principles.				
4. The summary presents an unbiased, objective interpretation of the article.				
5. The summary includes questions or confusing findings that were left unclear by the author.				
6. The summary accurately references the source.				
7. The summary is organized logically and easy to follow.				
8. The writer demonstrates correct usage, punctuation, spelling, and capitalization.				

Comments	Goals	Actions

Figure 7.50   Summarizing Scientific Articles (Performance List Rubric)

Understanding of Science (Performance List Rubric)

Name _____ Date _____ Course/Class _____

Task/Assignment _____

Performance Criteria	Assessment			
	Points	Self	Teacher	Other(s)
1. Given a scientific concept, I can list relevant supporting facts, explain them, show connections among them, and show connections of specific facts to the larger concept.				
2. Given a set of related concepts and facts on a specific scientific phenomenon, I can arrange these concepts and facts from most general to most specific and explain the relationships among them.				
3. Given an abstract scientific concept, I can develop mental and/or physical models to represent the concept.				
4. Using background information, prior knowledge, and personal observations, I am able to develop testable questions that can be answered through well-designed investigations.				
5. From testable questions I am able to develop and conduct well-designed investigations.				
6. Using experimental data I can analyze the data, both statistically and graphically, and interpret the data to develop inferences and conclusions.				

Figure 7.51   Understanding of Science (Performance List Rubric)

(Continued)

Understanding of Science (continued)

Performance Criteria	Assessment			
	Points	Self	Teacher	Other(s)
7. Using experimental data I can interpret patterns and/or trends and apply these patterns and trends when answering questions, solving problems, or predicting future events.				
8. Given sets of data on the same experiment from different experimental teams, I can compare the data sets and detect where either measurement or systematic errors may have occurred.				
9. I can interpret and evaluate the merits of different perspectives on a science question.				
10. Given a complex science and societal question/issue (such as water pollution) I can understand the varying points of view espoused by different groups and how decisions regarding the issue can affect each group in different ways.				
11. When reading science articles from newspapers, magazines, and other publications, I can explain the underlying scientific concepts for the issue(s) being presented.				
12. When reading science articles from newspapers, magazines, and other publications, I can interpret the author's intent (i.e., inform or persuade) and recognize any biases presented.				
13. As I plan and conduct scientific investigations and learn science concepts, I realize that what I know and can do is meager compared to the collective knowledge in the field of science.				

Comments	Goals	Actions

Web Site Design and Use (Performance List Rubric)

Name _____   Date _____   Course/Class _____

Task/Assignment _____

Performance Criteria	Assessment			
	Points	Self	Teacher	Other(s)
1. The purpose of the site is quite obvious.				
2. The layout is consistent throughout, extends the information naturally page-to-page, and is clear and user friendly.				
3. The home page has a well-labeled table of contents.				
4. The main path buttons look the same from page to page (Next/Previous/and Start Here) and are placed consistently.				
5. The main path buttons are relevant, clearly labeled, easy to navigate, and are not so numerous as to be overwhelming to the user.				
6. Each page contains a navigation bar that allows the user to return to the home page or main page of each section.				
7. Useful content is not more than three clicks away from the home page.				
8. Page titles are meaningful and are color or design coded to let the user know which section of the site they are in.				
9. Thumbnail graphics are used, can be quickly downloaded, are pertinent, and add instructional value.				

Figure 7.52   Web Site Design and Use (Performance List Rubric)

(Continued)

Web Site Design and Use (continued)

Performance Criteria	Assessment			
	Points	Self	Teacher	Other(s)
10. The text is easy to read and the language used is pertinent for the intended audience.				
11. The fonts are easy to read and point size varies appropriately for headings and text. The use of italics, bold, and underline contributes to the readability of the text.				
12. The site creates an excellent first impression and is engaging.				
13. Information contained within the site is accurate, complete, unique (not readily available elsewhere), and current.				
Optional The site provides user support.				
The site invites feedback and provides an email address for the contact person.				
Interactivity of the site adds to the instructional value.				
The site has a searchable index.				
The site announces the last time it was updated and links have been kept current.				

Comments	Goals	Actions

Writing an Article for Publication (Performance List Rubric)

Name _____ Date _____ Course/Class _____

Task/Assignment _____

Performance Criteria	Assessment			
	Points	Self	Teacher	Other(s)
1. The article follows any guidelines for publication that are issued by the publisher.				
2. The title or headline is appropriate for the intended audience and grabs their attention.				
3. The introductory paragraph clarifies the content of the article and provides the reader with some visual images of that which is to follow.				
4. Additional paragraphs build upon the introductory paragraph and completely and clearly develop the topic.				
5. The article "flows" from one paragraph to the next with no awkward transitions.				
6. The scientific content of the article is accurate.				
7. Visual aids (graphics, photographs, charts, graphs, etc.) enhance understanding of the text.				
8. Visuals are clearly titled, labeled, and referenced within the text.				
9. Sources for references are documented, appropriate to the topic, and current.				
10. Appropriate quotes are used to support points.				
11. Language is used purposefully, correctly, and is written in complete sentences.				

Comments	Goals	Actions

Figure 7.53   Writing an Article for Publication (Performance List Rubric)

Writing to Inform in Science (Performance List Rubric)

Name _____ Date _____ Course/Class _____

Task/Assignment _____

Performance Criteria	Assessment			
	Points	**Self**	**Teacher**	**Other(s)**
1. Accurate, specific, and purposeful scientific facts and concepts are extended and expanded to fully explain the topic.				
2. An organizational plan is established and consistently maintained.				
3. Scientific information that is relevant to the needs of the audience is used throughout the text.				
4. Scientific vocabulary and language choices enhance the text.				
5. Diagrams, pictures, and other graphics are of quality and add to the overall effectiveness of the text.				
6. There are no errors in the mechanics (spelling and grammar).				

Comments	Goals	Actions

Figure 7.54 Writing to Inform in Science (Performance List Rubric)

Writing to Inform in Science (Holistic Rubric)

Name _____ Date _____ Course/Class _____

Task/Assignment _____

Expert 4	Development: The writer provides accurate, specific, and purposeful scientific facts and concepts that are extended and expanded to fully explain the topic. Organization: The writer establishes an organizational plan and consistently maintains it. Audience: The writer provides scientific information relevant to the needs of the audience. Language: The writer consistently provides scientific vocabulary and language choices to enhance the text.
Proficient 3	Development: The writer provides scientific facts and concepts that adequately explain the topic with some extension of ideas. The information is usually accurate and purposeful. Organization: The writer establishes and maintains an organizational plan, but the plan may have some minor flaws. Audience: The writer provides information most of which is relevant to the needs of the audience. Language: The writer frequently provides scientific vocabulary and uses language choices to enhance the text.
Emergent 2	Development: The writer provides scientific facts and concepts that inadequately explain the topic. The information is sometimes inaccurate, general, or extraneous. Organization: The writer generally establishes and maintains an organizational plan. Audience: The writer provides some information relevant to the needs of the audience. Language: The writer sometimes provides scientific vocabulary and uses language choices to enhance the text.
Novice 1	Development: The writer provides insufficient scientific facts and concepts to explain the topic. The information provided may be vague or inaccurate. Organization: The writer either did not establish an organizational plan or, if an organizational plan is established, it is only minimally maintained. Audience: The writer did not provide information relevant to the needs of the audience. Language: The writer seldom, if ever, provides scientific vocabulary and uses language choices to enhance the text.

Comments	Goals	Actions

Figure 7.55   Writing to Inform in Science (Holistic Rubric)

Name _____ Date _____ Course/Class _____

Task/Assignment _____

Writing to Inform in Science (Analytic Rubric)

	Development	Organization	Audience	Language
Expert (4)	<u>Development</u>: The writer provides accurate, specific, and purposeful scientific facts and concepts that are extended and expanded to fully explain the topic. ☐	<u>Organization</u>: The writer establishes an organizational plan and consistently maintains it. ☐	<u>Audience</u>: The writer provides scientific information relevant to the needs of the audience. ☐	<u>Language</u>: The writer consistently provides scientific vocabulary and language choices to enhance the text. ☐
Proficient 3	<u>Development</u>: The writer provides scientific facts and concepts that adequately explain the topic with some extension of ideas. The information is usually accurate and purposeful. ☐	<u>Organization</u>: The writer establishes and maintains an organizational plan, but the plan may have some minor flaws. ☐	<u>Audience</u>: The writer provides information most of which is relevant to the needs of the audience. ☐	<u>Language</u>: The writer frequently provides scientific vocabulary and uses language choices to enhance the text. ☐
Emergent 2	<u>Development</u>: The writer provides scientific facts and concepts that inadequately explain the topic. The information is sometimes inaccurate, general, or extraneous. ☐	<u>Organization</u>: The writer generally establishes and maintains an organizational plan. ☐	<u>Audience</u>: The writer provides some information relevant to the needs of the audience. ☐	<u>Language</u>: The writer sometimes provides scientific vocabulary and uses language choices to enhance the text. ☐
Novice 1	<u>Development</u>: The writer provides insufficient scientific facts and concepts to explain the topic. The information provided may be vague or inaccurate. ☐	<u>Organization</u>: The writer either did not establish an organizational plan or, if an organizational plan is established, it is only minimally maintained. ☐	<u>Audience</u>: The writer did not provide information relevant to the needs of the audience. ☐	<u>Language</u>: The writer seldom, if ever, provides scientific vocabulary and uses language choices to enhance the text. ☐

Figure 7.56 Writing to Inform in Science (Analytic Rubric)

Writing Procedures for Conducting Scientific Investigations (Performance List Rubric)

Name _____ Date _____ Course/Class _____

Task/Assignment _____

Performance Criteria	Assessment			
	Points	Self	Teacher	Other(s)
1. A clear position is established that is fully supported or refuted with relevant, accurate scientific and/or personal information.				
2. A logical organizational plan for the text is established and consistently maintained.				
3. Scientific information that is relevant to the needs of the audience is used throughout the text.				
4. Scientific vocabulary and language choices enhance the position.				
5. Diagrams, pictures, and other graphics are of quality and add to the overall effectiveness of the position.				
6. There are no errors in the mechanics (spelling and grammar).				

Comments	Goals	Actions

Figure 7.57 Writing Procedures for Conducting Scientific Investigations (Performance List Rubric)

Name _____ Date _____ Course/Class _____

Task/Assignment _____

Writing to Persuade in Science (Analytic Rubric)

	Development	Organization	Audience	Language
Expert 4	Development: The writer identifies a clear position and fully supports or refutes that position with relevant, accurate scientific and/or personal information. ❑	Organization: The writer presents an organizational plan that is logical and consistently maintained. ❑	Audience: The writer effectively addresses the needs and characteristics of the identified audience. ❑	Language: The writer consistently uses relevant, scientific vocabulary and language choices to enhance the text. ❑
Proficient 3	Development: The writer identifies a clear position and partially supports or refutes that position with relevant, accurate scientific and/or personal information. ❑	Organization: The writer presents an organizational plan that is logical and maintained, but with minor flaws. ❑	Audience: The writer adequately addresses the needs and characteristics of the identified audience. ❑	Language: The writer frequently uses relevant, scientific vocabulary and language choices to enhance the text. ❑
Emergent 2	Development: The writer identifies a position, yet that position lacks clarity. The writer tries to support or refute that position with relevant, accurate scientific and/or personal information. ❑	Organization: The writer presents an organizational plan that is only generally maintained. ❑	Audience: The writer minimally addresses the needs and characteristics of the identified audience. ❑	Language: The writer sometimes uses scientific vocabulary and language choices to enhance the text. ❑
Novice 1	Development: The writer identifies an ambiguous position with little or no relevant, accurate scientific and/or personal information to support that position; or the writer fails to identify a position. ❑	Organization: The writer presents an argument that is illogical and/or minimally maintained. ❑	Audience: The writer does not address the needs and characteristics of the identified audience. ❑	Language: The writer seldom, if ever, uses scientific vocabulary and language choices to enhance the text. ❑

Figure 7.58   Writing to Persuade in Science (Analytic Rubric)

Writing for Personal Expression in Science (Analytic Rubric)

Name _____ Date _____ Course/Class _____

Task/Assignment _____

	Development	Organization	Audience	Language
Expert **4**	Development: The writer consistently develops science concepts, facts, ideas, conventions, and processes into a complete and well-developed whole. ☐	Organization: The writer purposefully orders scientific concepts, facts, and ideas. ☐	Audience: The writer fully anticipates and answers the audience's needs (audience may include self). ☐	Language: The writer consistently uses scientific vocabulary and language choices to enhance the text and in a manner appropriate to the literary form. ☐
Proficient **3**	Development: The writer partially develops science concepts, facts, ideas, conventions, and processes, but the response is not a complete, well-developed whole. ☐	Organization: The writer purposefully orders scientific concepts, facts, and ideas. ☐	Audience: The writer somewhat anticipates and answers the audience's needs (audience may include self). ☐	Language: The writer frequently uses scientific vocabulary and language choices to enhance the text and in a manner appropriate to the literary form. ☐
Emergent **2**	Development: The writer tries to develop science concepts, facts, ideas, conventions, and processes, but the response is not a well-developed whole and is not complete. ☐	Organization: The writer orders scientific concepts, facts, and ideas, but there are some interruptions in the flow of the piece. ☐	Audience: The writer attempts to anticipate and answer the audience's needs (audience may include self). ☐	Language: The writer sometimes uses scientific vocabulary and language choices to enhance the text and in a manner appropriate to the literary form. ☐
Novice **1**	Development: The writer has not developed science concepts, facts, ideas, conventions, and processes into a complete whole. ☐	Organization: The writer shows little purposeful ordering of scientific concepts, facts, and ideas. ☐	Audience: The writer has not anticipated and answered the audience's needs (audience may include self). ☐	Language: The writer seldom, if ever, uses scientific vocabulary and language choices to enhance the text and in a manner appropriate to the literary form. ☐

Figure 7.59 Writing For Personal Expression in Science (Analytic Rubric)

Writing Procedures for Conducting Scientific Investigations (Performance List Rubric)

Name _____ Date _____ Course/Class _____

Task/Assignment _____

Performance Criteria	Assessment			
	Points	**Self**	**Teacher**	**Other(s)**
1. The procedures are written in a logical order in a step-by-step format.				
2. Some type of numbering or labeling system is used to allow for easy referencing of a specific step.				
3. Scientific drawings or other graphics are added to enhance comprehension of the process/procedures.				
4. The written procedure includes a list of necessary materials and equipment.				
5. Any safety issues or precautions are listed.				
6. Both the dependent and independent variables have been identified within the steps.				
7. Control variables are listed within the steps.				
8. The steps test for the effects of only one independent variable at a time.				
9. The steps show that repeated trials were done.				
10. The steps are written clearly enough so that another person could repeat this experiment.				
11. Language is clear, grammatically correct, and written in complete sentences.				

Comments	Goals	Actions

Figure 7.60   Writing Procedures for Conducting Scientific Investigations (Performance List Rubric)

Written Interpretation of a Graph
(Performance List Rubric)

Name _____   Date _____   Course/Class _____

Task/Assignment _____

Performance Criteria	Assessment			
	Points	Self	Teacher	Other(s)
1. The title of the graph is restated in the summary.				
2. The dependent and independent variables are listed and their relationship is restated.				
3. Trends, patterns, and relationships displayed within the graph are described.				
4. The description of the displayed data is comprehensive and complete.				
5. A clear explanation of any discrepant data element is included.				
6. The interpretation focuses only on the actual data presented on the graph.				
7. Language is clear, grammatically correct, and written in complete sentences.				

Comments	Goals	Actions

Figure 7.61   Written Interpretation of a Graph (Performance List Rubric)

Web Sites for Rubrics in Science

The following Web sites contain a variety of assessment tools for different products and performances for many disciplines, not just for science. Educators K-12 will find these sites quite useful for ready-to-use, right-off-the-shelf assessment tools, or for tools that can be readily modified. Each Web site is followed by a brief description of the contents of that site.

Chicago Public Schools Rubric Bank

http://intranet.cps.k12.il.us/Assessments/Ideas_and_Rubrics/Rubric_Bank/rubric_bank.html

This site is very rich and contains an extensive bank of ready-to-use rubrics for Grades K-12 for many products and performances. Rubrics in reading, math, science, social studies, fine arts, speaking, and writing are available in PDF format that are easy to download. This site is among the best and should definitely be bookmarked.

The Staff Room for Ontario's K-12 Teachers

http://www.odyssey.on.ca/~elaine.coxon/rubrics.htm

This site has many content-specific examples ready to use! It is unique in that rubrics are available for content areas that are not routinely addressed. Also, links to other rubric Web sites are provided. There are tools (RubriStar, Rubric Generator, and Rubric Construction Set) available to personalize existing rubrics or create new rubrics. An excellent article on converting rubric scores into grades is provided. This site should definitely be bookmarked.

Kathy Schrock's Guide for Educators—Assessment Rubrics

http://school.discovery.com/schrockguide/assess.html

This outstanding site contains much information about performance assessment in general and has many links to other exceptional sites. Web page rubrics, general rubrics, articles on assessing student performance, ideas for portfolios, and suggestions for designing report cards are among the many features of this site. A searchable data bank of lesson plans in many content areas is available and is a very useful feature. This site should definitely be bookmarked and visited often.

PALS – Performance Assessment Links in Science

http://www.pals.sri.com

PALS is an online interactive resource bank for performance tasks and associated rubrics for science. It is standards-based (NSES) and continually updated. Each task includes student directions and response forms, administration procedures, scoring rubrics, examples of student work, and technical quality data calculated from field-testing. Online training packets have been created for some tasks. There is an online tour of the site, which is very helpful. This outstanding site should definitely be bookmarked.

Science Teacher Stuff
http://www.scienceteacherstuff.com/rubrics.html

The strength of this site is the many and varied links for rubrics, science resources, and teaching science that are provided.

Assessment Rubrics
http://www.what-is-the-speed-of-light.com/webquests/assessment-rubrics.html

This very useful site contains many links on introducing and creating rubrics. Included also are many examples of rubrics that can be used as they are or adapted for science performances.

Rubrician.com
http://www.rubrician.com/science.htm

This is an excellent site that should definitely be bookmarked and revisited often. This site works well for educators, teachers, students, parents, and evaluators. Included within this site are many links to science rubrics, such as physics lab rubrics, research cycle rubrics, science project checklists, scientific paper rubrics, science fair rubrics, primary school rubrics, and lab report rubric generator.

Exemplars
http://www.exemplars.com/rubrics/science_rubric.html

Exemplars provide examples of performance tasks and rubrics for science, mathematics, reading/writing/ research, and professional development. Spanish translations of the mathematics tasks will be available soon. Exemplar materials are linked to national standards and include performance tasks and their context, a subject-specific rubric, benchmark papers at four levels of performance, interdisciplinary links, and suggestions and time lines on how to implement the tasks. The bulk of these materials are available for purchase from Exemplars, but there are some free examples that can be downloaded.

Fairfax County Public Schools, Virginia
http://www.fcps.k12.va.us/DIS/OHSICS/forlang/PALS/rubrics

Presently, this site is devoted to performance assessment for language students (PALS). Both analytic and holistic rubrics are available for speaking, writing, interactive, presentational speaking, and Spanish for fluent speakers. Suggestions for scoring student work and converting the raw scores to percentage grades are included. This site is very worthwhile for teachers seeking examples of rubrics for language arts that could be modified for science.

Sackville High School
http://www.sackville.ednet.ns.ca/resource/teacher/sciencerubrics.html

Many links to science rubrics (K-12) are included among the contents of this Web site.

Write on Rhode Island

http://www.ri.net/WORI/

As the title suggests, this site is devoted to showcasing the writings of Rhode Island students. Rhode Island standards and various writing rubrics are found among the links. There are a number of helpful links for promoting writing, editing, and publishing of student work.

San Diego State University

http://webquest.sdsu.edu/rubrics/weblessons.htm

This site contains examples of rubrics, several articles on authentic assessment, suggestions on creating rubrics, a template for creating rubrics, and links to other rubrics Web sites.

http://webquest.sdsu/webquest.html

This is another San Diego State University site that contains many student-generated webquests in which the work of the students is generally assessed with rubrics. There are about 300 webquests for science K-Adult.

Rona's Ultimate Teachers Tools

http://www.theeducatorsnetwork.com/utt/rubricsgeneral.htm

This excellent site contains links to many examples of rubrics in a wide range of content areas.

Xerox Corporation—Science Rubrics

http://wwwtch.leusd.k12.ca.us/dscgi/ds.py/View/Collection-181

Rubrics for biology, integrated science, chemistry, physics, earth science, and investigation and experimentation are part of this site. Because the rubrics are correlated to specific science standards, they tend to be task specific and not generic.

Family Education Network

http://www.teachervision.com/lesson-plans/lesson-4521.html

This site contains a five-part series on how one teacher plans, revises, and implements rubrics in a variety of subjects. The site also includes lesson plans by topic, suggestions on how to assess student work, teacher tools, and many other worthwhile features.

References

Adams, C. M., & Callahan, C. M. (1995). The reliability and validity of a performance task for evaluating science process skills. *Gifted Child Quarterly, 39*(1), 14-20.

Andrade, H. (1997). Understanding rubrics. *Educational Leadership, 54,* 4.

Baron, J. B. (1991). Strategies for the development of effective performance exercises. *Applied Measurement in Education, 4,* 305-318.

Darling-Hammond, L. (1991). The implications of testing policy for quality and equity. *Phi Delta Kappan, 73*(3), 220-225.

Doran, R., Chan, F., & Tamir, P. (1998). *Science educator's guide to assessment.* Arlington, VA: National Science Teachers Association.

Educators in Connecticut's Pomperaug Regional School District 15. (1996). *A teacher's guide to performance-based learning and assessment.* Middlebury, CT: Association for Supervision and Curriculum Development.

Jamentz, K. (1994). Making sure that assessment improves instruction. *Educational Leadership, 51,* 6.

Lantz, H. (2001). *Developing performance-based units in science K-12.* Presentation at the National Science Teachers Convention, Baltimore, MD.

Lerner, L. S. (1998). State science standards: An appraisal of science standards in 36 states. *The Fordham Report.* Retrieved from http://www.edexcellence.net/library/lerner/gsbsteits.html.

Marzano, R. (2001). How and why standards can improve student achievement. *Educational Leadership, 59*(1), 14-18.

Marzano, R., Pickering, D., & McTighe, J. (1993). *Assessing student outcomes: Performance assessment using the dimensions of learning model.* Alexandria, VA: Association for Supervision and Curriculum Development.

McTighe, J. (1999). *Developing performance tasks: Tools and templates for designers.* Columbia: Maryland Assessment Consortium.

National Research Council. (1996). National science education standards. Washington, DC: National Academy Press.

Popham, W. J. (1997). What's wrong—and what's right—with rubrics? *Educational Leadership, 55,* 2.

Project 2061, American Association for the Advancement of Science. (1993). *Benchmarks for science literacy.* New York: Oxford University Press.

Silver, H., Strong, R., & Perini, M. (2000). *So each may learn: Integrating learning styles and multiple intelligences.* Alexandria, VA: Association for Supervision and Curriculum Development.

Treagust, D., Jacobowitz, R., Gallagher, J., & Parker, J. (2003). *Embed assessment in your teaching.* Arlington, VA: National Science Teachers Association.

Wiggins, G., & McTighe, J. (1998). *Understanding by design.* Alexandria, VA: Association for Supervision and Curriculum Development.

Index